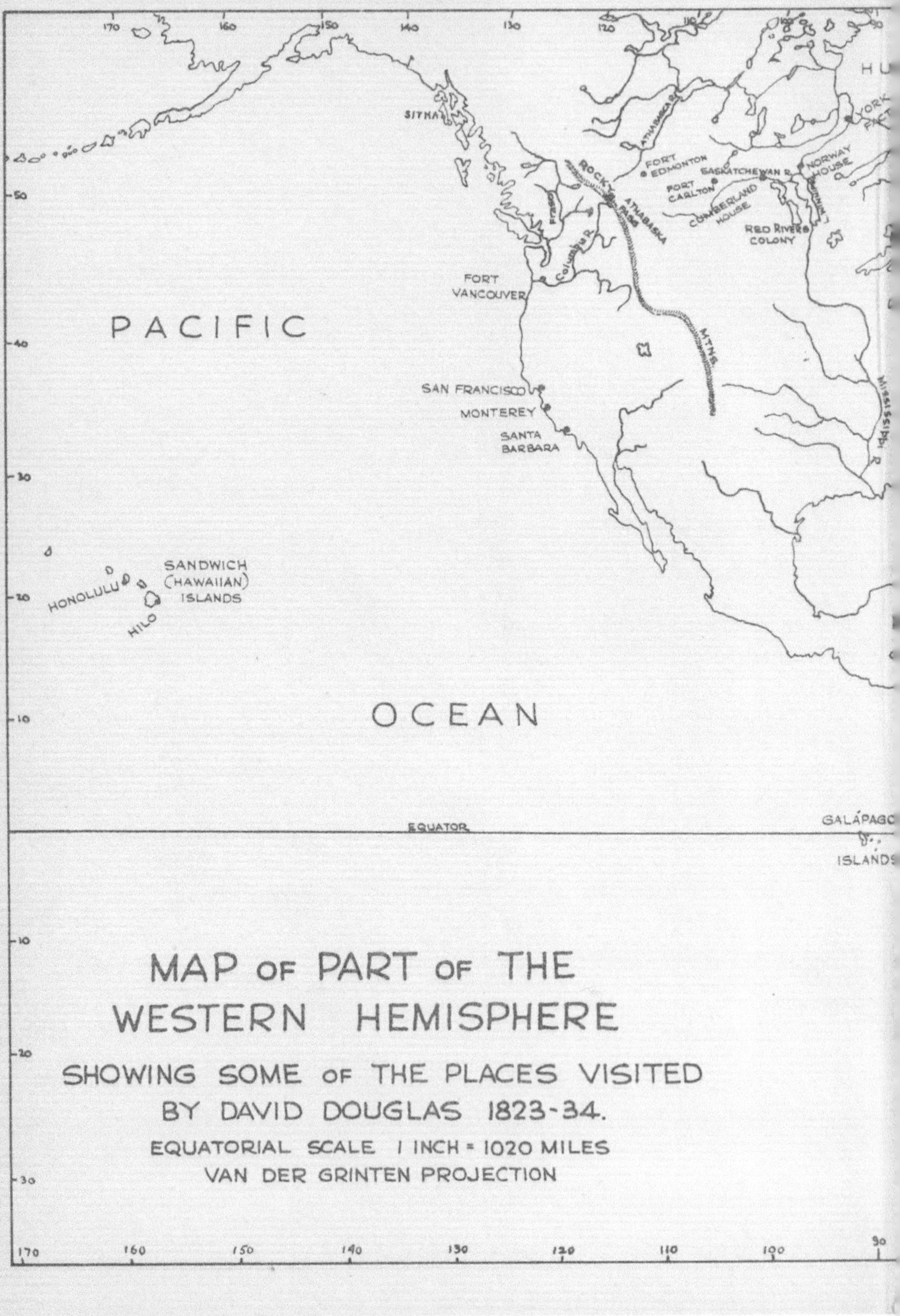
MAP OF PART OF THE WESTERN HEMISPHERE
SHOWING SOME OF THE PLACES VISITED BY DAVID DOUGLAS 1823-34.
EQUATORIAL SCALE 1 INCH = 1020 MILES
VAN DER GRINTEN PROJECTION
PACIFIC
OCEAN
SITKA
FORT EDMONTON
FORT CARLTON
SASKATCHEWAN R.
CUMBERLAND HOUSE
NORWAY HOUSE
RED RIVER COLONY
ROCKY MTNS.
ATHABASKA
ATHABASKA R.
FRASER
Columbia R.
FORT VANCOUVER
SAN FRANCISCO
MONTEREY
SANTA BARBARA
SANDWICH (HAWAIIAN) ISLANDS
HONOLULU
HILO
EQUATOR
GALÁPAGO
ISLANDS
Mississippi R.
170
160
150
140
130
120
110
100
90
50
40
30
20
10

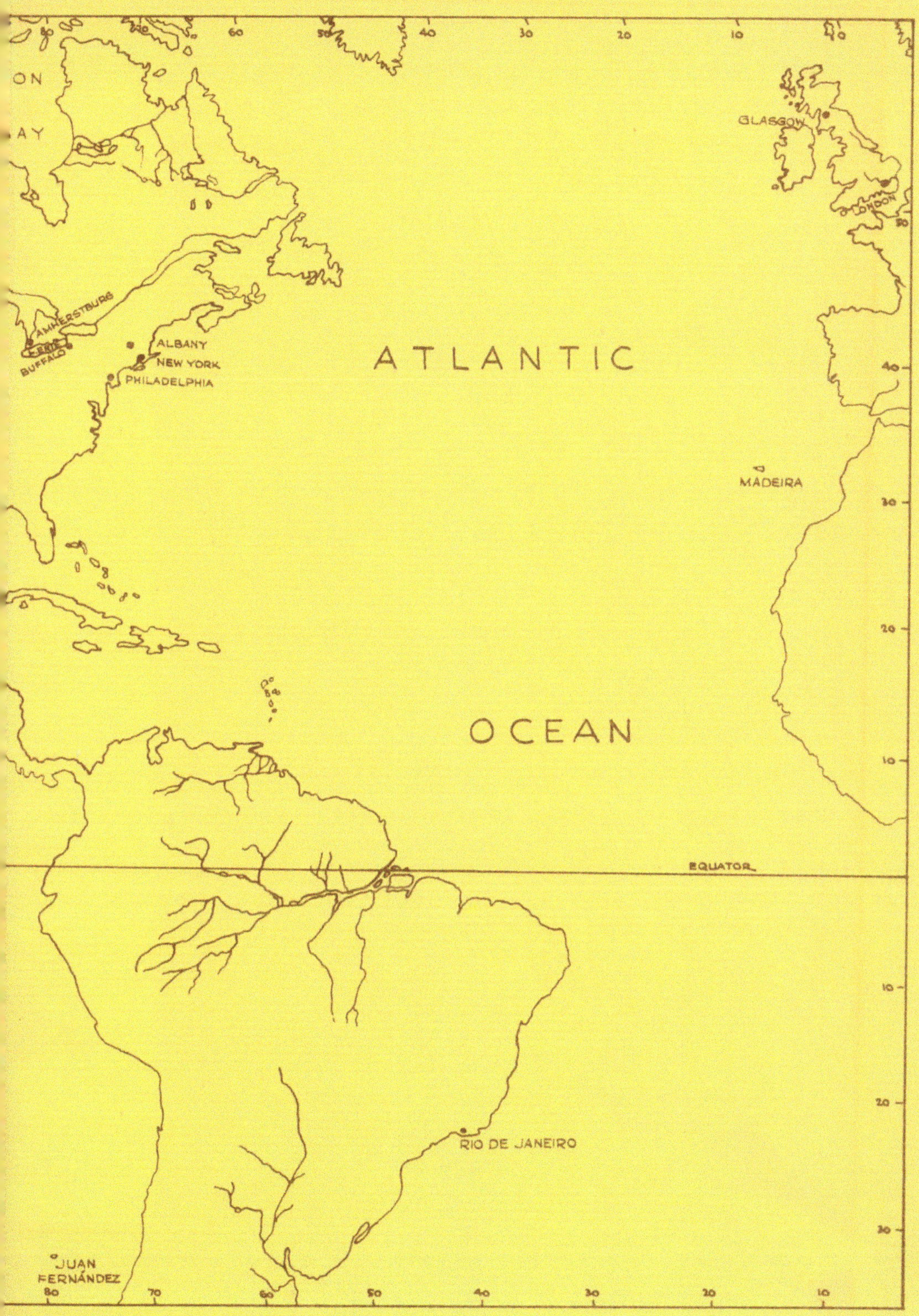

ON
AY
AMHERSTBURG
ERIE
BUFFALO
ALBANY
NEW YORK
PHILADELPHIA
GLASGOW
LONDON
ATLANTIC
OCEAN
MADEIRA
EQUATOR
RIO DE JANEIRO
JUAN
FERNÁNDEZ
60
50
40
30
20
10
0
80
70
60
50
40
30
20
10
50
40
30
20
10
10
20
30

DOUGLAS OF THE FIR

LONDON : GEOFFREY CUMBERLEGE
OXFORD UNIVERSITY PRESS

DAVID DOUGLAS *from a pencil drawing by his niece, Miss Atkinson, 1829*

DOUGLAS OF THE FIR

A BIOGRAPHY OF
DAVID DOUGLAS
BOTANIST

ATHELSTAN GEORGE HARVEY

HARVARD UNIVERSITY PRESS
CAMBRIDGE, MASSACHUSETTS
1947

Printed in the United States of America

PREFACE

I BECAME INTERESTED in my subject through the casual remark of a friend. A few years ago while chatting with the late Mr. Frank Stevens Hall, Director of the Museum of the Eastern Washington Historical Society at Spokane, I learned that he was preparing an article on birds collected in the Pacific Northwest by a David Douglas. Upon inquiring who this man was, I was told something of his life and that the Douglas fir tree was named after him. Up to that time I had understood that the tree had been named for Sir James Douglas, early governor of British Columbia—an erroneous impression all too prevalent in this province. My interest in the intrepid young explorer being aroused, I read his journals and everything else I could find about him. I hunted up material that had not been published and other material that had been published and forgotten.

During my quest it occurred to me that other persons might be interested, and that at all events David Douglas deserved to be better known; so in my spare time I proceeded to obtain all the information about him that I could and to arrange it in biographical form. By correspondence or personal investigation bits of information were obtained from various persons and places. Archives and libraries contributed, as far removed from one another as London and Honolulu. This book is the result.

I desire to express my thanks to the many who in one way or another have assisted me. To name them all would make too long a list. Some are mentioned in the text or the notes. Others to whom especial thanks are due are: Dr. W. Kaye Lamb, Librarian of the University of British Columbia; Mr. E. S. Robinson, City Librarian, Mr. Robert Allison Hood, Mr. Noel Robinson, and Mr. Edward T. Oliver, all of Vancouver, British

Columbia; Mr. W. A. Newcombe of Victoria, British Columbia; Mr. L. W. Bryan, Associate Forester, Hilo, Hawaii; and Mr. W. A. McAdam, Agent General for British Columbia, London. Nor can I omit mention of my wife, whose advice and encouragement have been a continual help.

Information obtained from the Archives of the Hudson's Bay Company is published with permission of the Company's Governor and Committee.

In considering statements of present facts the reader should bear in mind that the book has been in course of preparation for some years.

A. G. H.

Vancouver, British Columbia
May, 1947

CONTENTS

BIOGRAPHICAL SUMMARY ix

PROLOGUE 3

1. SCOTLAND 7

2. THE HORTICULTURAL SOCIETY 17

3. UNITED STATES AND UPPER CANADA 22

4. FIRST GREAT ADVENTURE 37

5. COLUMBIA RIVER REGION 49

6. WITH CHIEF THA-A-MUXI 62

7. UP THE COLUMBIA TO KETTLE FALLS 68

8. THE BLUE MOUNTAINS 79

9. IN SEARCH OF THE SUGAR PINE 91

10. HOMEWARD BY HUDSON'S BAY EXPRESS 107

11. PUTTING MOUNTAINS ON THE MAP 130

12. FAME AND ITS AFTERMATH 144

13. SECOND GREAT ADVENTURE 157

14. CALIFORNIA 174

15. THIRD VISIT TO THE COLUMBIA 190

16. ILL-FATED JOURNEY. 197

17. FLAKES OF GOLD 209

18. HAWAII AND ITS VOLCANOES 212

19. LAST JOURNEY 232

20. IN TRIBUTE 238

APPENDICES

A. ACCIDENTAL DEATH OR MURDER? 251

B. PLANTS INTRODUCED BY DOUGLAS 254

BIBLIOGRAPHY 261

INDEX 277

ILLUSTRATIONS

DAVID DOUGLAS. *From a pencil drawing by his niece, Miss Atkinson, 1829* . . . . Frontispiece

WILLIAM JACKSON HOOKER, *Professor of Botany, University of Glasgow, 1820 to 1840; Director of Kew Gardens, 1841 to 1865* . . . 14

DOUGLAS FIR TREES. *Leonard Frank Photos, Vancouver, British Columbia* 86

FORT VANCOUVER. *Provincial Archives, Victoria, British Columbia* . 87

JASPER HOUSE. *Provincial Archives, Victoria, British Columbia* . . 87

ATHABASKA PASS, *showing Mount Brown and the Committee's Punch Bowl. Hydrographic Service, Ottawa, Ontario* 118

MOUNT HOOD. *Oregon State Highway Commission, Salem, Oregon* . 119

MAPS

PART OF THE WESTERN HEMISPHERE, *showing some of the places visited by David Douglas, 1823 to 1834* . . . . End papers

PART OF THE PACIFIC NORTHWEST *as Douglas found it* . . . 50

MOUNT BROWN AND MOUNT HOOKER, *as shown on the map published in W. J. Hooker's* Flora Boreali-Americana *(London, 1840)* . . 133

PART OF THE SANDWICH (HAWAIIAN) ISLANDS 214

DAVID DOUGLAS

BIOGRAPHICAL SUMMARY

1799 Born June 25th at Scone, Perthshire, Scotland.

1810 or 1811. Apprenticed at Earl of Mansfield's gardens, Scone, for seven years.

1818 To Sir Robert Preston's garden, Valleyfield, near Culross, for two years.

1820 To the Botanical Garden, Glasgow.

1823 In the spring, enters the employ of the Horticultural Society of London. On June 6th leaves Liverpool on a journey to eastern America, arriving in New York on August 3rd. After four months' travel in the United States and Upper Canada, leaves New York on December 12th for London.

1824 Arrives in London on January 9th. On July 25th leaves London for northwest America, via Cape Horn.

1825 On April 7th arrives at the mouth of the Columbia River. Reaches Fort Vancouver on April 20th.

1827 After two years' travel in northwest America, leaves Fort Vancouver on March 20th for England, via Hudson Bay. Reaches York Factory on August 28th, and on October 11th arrives at Portsmouth.

1829 Following two years' sojourn in England, on October 31st sets out again for northwest America, via Cape Horn and the Sandwich Islands.

1830 Reaches the mouth of the Columbia on June 3rd. Spends the summer and fall making journeys in the interior, east and south. On December 10th leaves the Columbia for California, arriving at Monterey on December 22nd.

1832 After nineteen months' travel in California, departs on August 18th for the Columbia, via the Sandwich Islands. Stops at Honolulu from September 7th to 12th. Resigns his position with the Horticultural Society of London on September 9th. Reaches the Columbia on October 14th.

1833 February to March, makes journey to Puget Sound. Leaves Fort Vancouver on March 19th for journey to New Caledonia, returning in July. On October 18th leaves the Columbia for the Sandwich Islands, via California. Stops at San Francisco from November 4th to 29th, then sails for Honolulu, arriving December 23rd. On the 27th leaves Honolulu for Byron's Bay (Hilo), Hawaii.

1834 Reaches Byron's Bay on January 2nd, and spends three months in travels on the island of Hawaii. Leaves for Honolulu on March 29th. Returns to Hawaii again on July 9th, and is killed there on the 12th. Buried at Honolulu on August 4th.

PROLOGUE

PROLOGUE

It is late afternoon of a bright April day in the year 1825. The weather-beaten three-masted sailing vessel *William and Ann* has just finished crossing a treacherous sandbar and is cautiously entering the mouth of the little-known Columbia River. From the deck come sounds of merriment, in which all on board seem to be participating. Having sailed from England, around Cape Horn, they are joyful that the last danger of their perilous voyage of eight and a half months is safely passed.

Standing at the rail, one of the passengers shows his joy in a distinctive manner. A sturdy, fair-faced young Scotsman, he is staring intently across the water, while with characteristic eagerness he studies the forest-covered shore. It is the trees that have his attention. One resembles the eastern hemlock; another looks like a balsam fir. A third, which he cannot surely identify, has deep yellow-green foliage similar to that of the hemlock, only more beautiful; and it is a much statelier tree. Its many spreading branches—the lower ones drooping slightly, the higher ones curving upward—with their gracefully hanging branchlets, its symmetrical shape, and its towering pyramidical crown excite his curiosity and admiration.

Little does the weary voyager imagine that this noble tree, destined to become the world's greatest producer of structural timber, will be named in his honor, and be known as the "Douglas fir."

He retires early to his berth for a good night's rest—the first in several weeks. Tomorrow he will land. The botanizing to which he has been looking forward all these months, and which has already begun from the deck of the ship, will proceed with earnest, almost desperate, zeal. The territory is a primeval wilderness inhabited only by aborigines and a few fur hunters. Incredible perils and hardships must be faced. Nevertheless, dauntless with devotion and enthusiasm, at times almost oblivious of danger, he will systematically comb it for botanical specimens: trees, shrubs, flowers, plants, and their seeds. Other branches of natural history—birds, animals, minerals—will not be overlooked. His interest will range from tiny plant to stupendous mountain. His travels will extend throughout the immense Columbia River territory and across the continent to Hudson Bay.

Success will attend his efforts. Dispatching seeds and specimens to England, he will introduce to the world not only the giant fir he has observed from the deck of the *William and Ann*, but also several other trees and such a host of new flowers and plants as will startle the botanical world. Honors will be heaped upon him.

He will return to the Columbia. California, too, will be combed; then Hawaii. Here his all-too-brief career will end in tragedy. But in the short span of his life, and in his particular sphere, he will have accomplished more than any man of his time. It will be said of him that "there is scarcely a spot deserving the name of a garden, either in Europe or the United States, in which some of the discoveries of Douglas do not form the chief attraction; to no single individual is modern horticulture more indebted."

DOUGLAS OF THE FIR

1

SCOTLAND

OLD SCONE is a place of the past. The grandeur and glory of its ancient days have disappeared. No more is it the capital of Scotland, coronation place of kings, meeting place of parliaments. No longer is it the "Royal City of Scone." Gone, indeed, is the old village itself. Only the palace remains.

A century and a half ago the village was still standing. It was a cluster of dwellings behind the palace, on the site of the present driveway. Toward Perth ran the Chantorgait, where in olden times the village chantor used to sing the evening hymn. Along this street was the Wilderness Den, a little depression in the ground, up the steep slopes of which the village penitents once were forced to climb on bare knees in penance for their sins. Nearby was the Gallows Knowe, a tiny knoll upon which it had been the custom to mete out the worst punishment of all.

One of the humble cottages was occupied by the village stonemason, John Douglas. There on June 25, 1799,[1] his wife, Jean Drummond, gave birth to a son whom they named David.

The father was fairly well educated for one of his station in life, and was reserved and dignified in manner. He had a reputation for first-class workmanship and this often brought him special jobs. He put up most of the tombstones in the churchyards of

[1] John Douglas, "Biographical Notes of David Douglas," MS., Kew Gardens; David Douglas, *Journal Kept by David Douglas during His Travels in North America, 1823-1827* (London, 1914), pp. 3, 152; W. J. Hooker, "A Brief Memoir of the Life of Mr. David Douglas, with Extracts from His Letters," *Companion to the Botanical Magazine*, London, II (1836), 79 (article reprinted in the *Hawaiian Spectator*, Honolulu, II, 1839, and in the *Oregon Historical Quarterly*, Portland, Ore., V and VI, 1904 and 1905. Subsequent references will be to the *Oregon Historical Quarterly*, abbreviated *OHQ*. This reference is to *OHQ*, V, 223). Some writers erroneously give the year as 1798.

the neighborhood. He was called in by the gentry living far and near to set up their kitchen ranges in the chimney places and to cure their smoky flues. Although these occasions might have been used as opportunities for substantially adding to his income, his extreme honesty led him to charge little more than his usual wage; and because of this some of his neighbors considered him lacking in worldly wisdom. Nevertheless, like many other Scottish parents, by frugality and self-sacrifice he managed to bring up his family comfortably and to educate them reasonably well.

There were six children: three boys and three girls. John, the eldest, became apprenticed to William Atkinson, the eminent architect, and afterwards was clerk of works to the Duke of Buccleuch. George, youngest of the family, became cashier to a pottery works in Staffordshire.[2] Little is known about the girls. One married an Atkinson, two died unmarried. No descendant of any of the children is now alive.[3]

David's education began at an early age. When about three years old he was sent to the village school, where he soon made a name for himself, but not as a student. The teacher, a "good old dame," we are told, "gentle of heart nor knowing well to rule," found herself mastered by her sturdy, high-spirited little pupil, who showed his dislike for confinement and discipline by playing truant and otherwise defying her authority. Another account gives the teacher the somewhat different character of an "inflexible disciplinarian," but agrees on the character of the child. The result was that the continual clashes, aggravated by some special act of self-will, led to his removal from the school.

No doubt the child's stubbornness was due largely to heredity and parental influence. The father, as we have seen, had great independence of mind and strength of character. Then, too, there was the religious atmosphere in which he was brought up.

[2] Letter, April 21, 1936, Miss M. B. Tennant, Longport, Stoke-on-Trent, Staffordshire, to the author.

[3] Letter, August 15, 1937, Miss M. L. McFarlane, Scone, to the author.

The family were not of the established faith, that of the Church of Scotland, but showed their independence by attending a church of dissenting Presbyterians,[4] which afterwards was included in the United Presbyterian Church.

Just how old Douglas was when he left the Scone school is not known. It may not have been until after the family had moved to New Scone, two miles away, which took place when he was six or seven. The whole village was moved as part of a larger plan for rebuilding the palace and enlarging its grounds. Even the parish church was moved. The stones were numbered and taken down one by one, then carted to the present site, where the church was reërected.

Whether it was at the old location or the new that the family was living, David's schooling did not end. Some three miles away (from either place) at Kinnoul on the outskirts of Perth was another school; and, unlike that at Scone, it had a male teacher. So there the young rebel was sent, in the hope that the change from schoolmistress to schoolmaster would be for the better.

It was—to some extent. He took to book learning slowly. He preferred nature and the fields. To wander over meadow or through woodland, collecting flowers and plants; to explore the haunts and observe the peculiarities of bird, beast, and fish; to climb trees for birds' nests and carry off their spotted treasures —these were his delights. Schoolmaster Wilson often had to punish him for not knowing his lesson or for being late or playing truant. But punishment of that sort did little good. The taws, that stinging, cutting, blistering instrument of chastisement so dreaded by Scottish schoolboys of his day, held no terrors for him. He could suffer it and soon forget. The punishment that he dreaded was to be kept in after school. It meant less time for his outdoor enjoyment.

"His contempt of the schoolmaster's thong," says a friend of the family (Fish), "and his carelessness about those difficulties

[4] Letter, June 1, 1936, Miss M. L. McFarlane, Scone, to the author.

and hardships which would have weighed hard with other boys, were budding into that strong-minded, self-dependent heroism which enabled him afterwards to dare and do so much for the advantage of natural history."

The six-mile walk to and from school may have been designed to exercise a sobering influence on the high-spirited boy. Instead, it had the opposite effect: it strengthened and encouraged him in his natural bent. It gave greater opportunity for observation and study of the plant and animal life he loved. This he could do along the way as he went back and forth on the daily journey. And behind the school there was Kinnoul Hill, with its forested or heather-clad slopes, convenient for play and nature study at the noon hour.

At fishing he became successful, alike with proper tackle or makeshifts. When he could not obtain the former he resorted to a willow pole, a piece of string, and a crooked pin, and often had satisfactory results.

He collected not only birds' eggs, but also the birds themselves, "all sorts," we are told by Hooker, "though he often found it difficult to maintain some of these favourites, especially the hawks and owls. For the sake of feeding a nest of the latter, the poor boy, after exhausting all his skill in catching mice and small birds, used frequently to spend the daily penny with which he should have procured bread for his own lunch, in buying bullock's liver for his owlets, though a walk of six miles to and from school might well have sharpened his youthful appetite."

Indifferent as David was to school work, he was no dullard. What he lost by scholastic indifference was offset by the gain from his outdoor activities. The search for wild life, and its observation and study, sharpened his eye, quickened his perception, and broadened his mind.

"He had at all times a very inquisitive disposition," says his older brother, "and was not satisfied with merely seeing a new thing; but must know where it came from, and how it was procured. As an instance of this I remember one day in passing a Tobacconist's shop in Perth he noticed the figure of a black man preparing tobacco, which so aroused his attention that he could not rest satisfied until he had learned its whole history."[5]

Another characteristic appearing early in life was his interest in foreign countries. He delighted in reading and talking about the experiences of travelers abroad. Books of adventure thrilled him, and of these *Sinbad the Sailor* and *Robinson Crusoe* were his favorites.

He continued at the Kinnoul school until he was ten or eleven. At this early age his formal education came to an end and he was put to work. The interest he had shown in plants won him an apprenticeship in the Scone Palace gardens, which were under the superintendence of head gardener William Beattie, a friend of his father's. The palace, just rebuilt by its owner, the Earl of Mansfield, at a cost of £70,000 and made one of the finest castellated mansions in Scotland (with 125 rooms, 90 of them bedrooms), was matched by the magnificence of its grounds. These included a thousand acres of forests, fields, parklands, and gardens. Like the palace, they too had their mementos of royal visits. There was the sycamore tree planted by unfortunate Queen Mary; and nearby were the oak and the sycamore set out by the trowel of King James VI.

The gardens of the palace were not unfamiliar to the boy; he had been born and brought up close to them, almost in them. Yet they were not for him by right. While the child of a villager might admire at a distance, or even enter circumspectly, the grounds were not his to enjoy. That privilege was for the occupants of the palace. Now, however, as a worker in them, he was not only to savor of their beauty, but was to become familiar with the whole course of plant production and growth throughout the seasons.

Beginning in the flower garden under young Mr. McGillivray, he found the work entirely to his liking—so different from the

[5] John Douglas, "Biographical Notes."

hated school work. His stubbornness and disobedience disappeared and were replaced by willingness, diligence, and zeal. Now his real education began. Entering wholeheartedly upon his work, he quickly learned the names of the various species of flowers and plants and mastered the rudiments of botany, so that he soon knew as much as McGillivray. Thus began the orderly development of the taste for botanical pursuits which dominated his life.

The study of plants, hitherto merely a hobby pursued irregularly and furtively under the ban of authority, had become a vocation performed openly under the auspices of nobility and the guidance of a skilled staff. At last he could feel he was on the right track, and he showed his appreciation of the opportunity by the energetic, intelligent manner in which he did his work.

Mr. Beattie took to him greatly. He was not the only boy in the garden, and his independent and mischievous disposition now resulted in clashes, not with his instructors as heretofore, but with the other lads. Upon their complaining about him to their master, Beattie showed his regard for David's worth and ability and forestalled further complaints by answering bluntly: "I like a devil better than a dult."

Always fond of books on travel, he now widened his field to embrace works on natural history. These he studied thoroughly, making extracts of portions which took his fancy and memorizing them. He read everything on the subject that he could lay his hands on, often borrowing from friends and acquaintances. On one occasion, having asked an old friend (Mr. Scott) if he could lend him anything suitable, he was surprised to receive a Bible, accompanied by the kind admonition: "There, David, I cannot recommend a better or more important book for your perusal."

While the winter nights were given to reading and study, the summer evenings were spent in short botanical trips, often in company with other lads similarly inclined. The living plants that he collected were put in a small garden at his home; the other specimens were mounted on paper. It is noteworthy that these trips were never taken on the Sabbath, his father having strictly prohibited excursions on that day; and this rule the boy obeyed faithfully.

The excellence of the gardens at Scone Palace was well known to nurserymen and gardeners, and there were many visitors. Frequent among them were Messrs. R. and J. Brown of the Perth Nursery (one of the best in Scotland, distinguished for the introduction and distribution of new and useful plants). The Browns were enthusiasts and devoted part of each summer to botanizing in the Highlands. Making their acquaintance through Mr. Beattie, young Douglas learned of these excursions with much interest. Indeed, it is probable that the Browns' accounts of their adventures and of the romantic scenery of the places they had visited in search of plants led him to resolve secretly that some day he would follow their example.

Having completed the usual term in the ornamental department he went on to the forcing and kitchen garden, where he took the same lively and diligent interest as in the former. For one with a taste for botany the homely vegetable has its attractions as well as the gorgeous flower, and to him each made its appeal. There was a change also in his manuals of study; *Lee's Introduction to Botany* and *Donn's Catalogue,* his former textbooks, were laid aside, and *Nicol's Gardener's Calendar* was taken in their place. The more modern books and periodicals on horticulture had not yet appeared, so that the means of learning the theoretical side of the business were very limited. Nevertheless what he lacked in theory he more than made up in practice; and it is said that when his seven years' apprenticeship was finished his ability was such that he needed only age and experience in managing men to qualify him for "a situation of the first importance."

An eighteen-year-old boy could hardly be expected to take charge of a staff of dour Scottish gardeners. There were still some rungs of the ladder to climb. So, after spending the winter of 1817-18 at a private school in Perth, with special attention to arithmetic, he secured a position for which he was suited. This was an appointment (made upon Mr. Beattie's recommendation) under Alexander Stewart, gardener to Sir Robert Preston at Valleyfield, near Dunfermline. Here was a grand selection of plants, particularly of exotic varieties—one of the best private collections in Scotland—and young Douglas was delighted. A new idea came to him: Instead of being a gardener why not become a botanist? Why not get an intimate knowledge of these lovely foreign plants? Why not hunt for them in their native habitats?

He was encouraged in this idea by being given access to Sir Robert's botanical library, a privilege of the utmost value to one endowed with such unusual mental faculties but hampered by financial circumstances. Further encouragement and reward for his earnest work were received in his being made foreman to Mr. Stewart.

After about two years at Valleyfield (the last year acting as foreman) he obtained a place on the staff of the famous botanical garden at Glasgow. To this occupation, the best that he had attained so far, he devoted himself with his usual earnestness and enthusiasm, and he took full advantage of its wider opportunities. He was no mere employee, but an ardent student working among the things he loved.

Quickly winning the friendship and esteem of the curator, Stewart Murray, he went on to form another friendship, one which helped him tremendously throughout his life. In the hall of the garden lectures on botany were given by Dr. Hooker, Regius Professor of Botany at Glasgow University,[6] and to these

[6] William Jackson Hooker (1785-1865) botanized in Scotland, Iceland, continental Europe; Regius Professor of Botany, Glasgow University, 1820-1841; Director, Royal Botanic Gardens, Kew, 1841-1865; author of several botanical works, including *Flora Scotica, Flora Boreali-Americana,* and *British Flora;* LL.D. Glasgow, D. C. L. Oxon., Knight of Hanover; liberally assisted young botanists; maintained friendly relations with Colonial Office and other govern-

WILLIAM JACKSON HOOKER *Professor of Botany, University of Glasgow, 1820 to 1840; Director of Kew Gardens, 1841 to 1865*

lectures young Douglas was not slow in finding his way. Strange to say, Hooker had never taught before accepting the Glasgow professorship, nor even heard a course of lectures, yet his success was phenomenal, owing to the thoroughness and zeal of his teaching and to his kindly personality. He was an instructor whose enthusiasm overflowed into his students and he was a moving power with many of them. He had the art of making them love the science he taught. So popular were his lectures that they were attended not only by students of medicine, for whom the course was primarily designed, but also by gentlemen from the city and military officers from the barracks.[7]

"The garden," we are told by one who knew it, "was then most unfavourably situated in the midst of reeking chimneys and murky factories.[8] The lecture-room was a small, dingy building in its centre—only different from a dog-kennel in being upstairs. Since that time, both the university and the garden have been removed to more favourable localities. Yet in that old building some noble work was done. There Hooker gave those lectures which established his name as a prince in Botanical Science. These lectures Douglas attended while engaged there. His soul was set on fire with zeal. Not any of the medical students, who attended as part of their curriculum, made better progress, or attracted more the attention of the warm-hearted Professor." [9]

At that period educational advantages for a boy of humble birth were few, and although David was naturally bright and

---

ment offices, and persuaded them to do much to advance botanical science. Father of Sir Joseph Dalton Hooker, also a famous botanist.

[7] J. D. Hooker, "A Sketch of the Life and Labours of Sir William Jackson Hooker," *Annals of Botany*, London, XVI (1902), xxviii-xxx; F. O. Bower, "Sir William Hooker," in *Makers of British Botany*, F. W. Oliver, ed. (Cambridge, England, 1913), p. 130.

[8] The garden consisted of about six Scottish or eight English acres, and occupied the area which now is along both sides of Blythswood Street between Argyle and St. Vincent Streets (*Companion to the Glasgow Botanic Garden*, Glasgow, c. 1818, pp. 8, 9).

[9] Rev. Thomas Somerville, "An Early Hero of the Pacific; Douglas, the Botanist," *Overland Monthly*, San Francisco, VII (August, 1871), 106-107.

was determined to get ahead (and there were many instances of successful achievement through perseverance and determination long before the days of free education), yet he was fortunate in coming into contact with a man of such ability and culture as Hooker. The intimate friendship which developed was undoubtedly the chief benefit derived by him from his period of service at Glasgow. It deepened and strengthened as the years went by, and was destined to be ended only by his tragic death in the wilds of Hawaii at the age of thirty-five.

As an early result of this friendship he was chosen to accompany the professor on some botanical excursions to the Scottish Highlands and Islands, and indeed became his favorite companion.[10] The hope of imitating the Browns of Perth had come true. The close comradeship of these journeys—the tramping, searching, studying, and collecting from sunrise to twilight, over heathered moor, along stream or shimmering loch, up glen or mountainside or over bleak, wind-swept island, regaling themselves now and then with bannocks and milk or a snatch of bread and cheese and a sup of ale, sleeping for the few hours of darkness on the floor of some outlying tavern or shepherd's shieling or in some hay loft—gave Hooker an excellent opportunity of observing the young man closely and gauging his character and his gifts.

What he discovered in this way more than satisfied him. He was convinced that this was no ordinary student, and that here was a youth who deserved a wider field of activity. Says Hooker: "His great activity, undaunted courage, singular abstemiousness, and energetic zeal at once pointed him out as an individual eminently calculated to do himself credit as a scientific traveler."

[10] *Ibid.*

2

THE HORTICULTURAL SOCIETY

And God said, Behold, I have given you every herb bearing seed, which is upon the face of all the earth, and every tree, in the which is the fruit of a tree yielding seed; to you it shall be for meat.

GENESIS, I:29.

THE INTRODUCTION of strange plants from foreign lands is one of the oldest activities of mankind. It arose out of the elemental physical needs for more and better food, clothing, and shelter. With the advance of civilization other motives came into play: the desire for the beautiful and the urge for scientific study. It is nearly four thousand years since Queen Hatshepsut of Egypt sent an expedition abroad to collect trees. One of her successors, King Thotmes III, sent a botanical expedition to Syria about 1500 B.C. and was so pleased with the result that he had the iris, heather, lotus, and other flowers sculptured on the temple walls at Karnak, where they may still be seen.

When peace came to Europe after the Napoleonic wars the age-long yearning become stronger than ever. In particular demand were species from distant lands, because of their oddity and vivid colors. They made attractive features in ornamental gardens. "Distinguished persons," we are told, "would throw themselves almost into a frenzy when the rumour went abroad that some new or rare specimen was about to appear on the market." [1] Thus encouragement was given to the sending of plant-collecting expeditions to all parts of the world. Chosen for the work were young men well trained in botany and willing to face great hardship and risk in order that the desires of gar-

[1] Violet R. Markham, *Paxton and the Bachelor Duke* (London, 1935), p. 53.

den lovers and scientists might be satisfied. Such was the work in which David Douglas was to engage.

One of the chief promoters of these expeditions was the Horticultural Society of London. This organization (now the Royal Horticultural Society) was founded in 1804 by John Wedgwood (of the famous pottery family) and chartered in 1809, its general purpose being to encourage and elevate the study and practice of horticulture. With His Majesty King George IV as patron, it was supported by the leading men of the kingdom, and in 1823 had about 1500 members ("fellows"), besides honorary members, foreign members, corresponding members (some at home, others abroad), and practical gardeners. Among the foreign members were the Emperor of Russia and the kings of Denmark, Bavaria, and the Netherlands. The president was Thomas Andrew Knight, a country gentleman and a writer on vegetable physiology. The secretary was Joseph Sabine, a lawyer who had turned from law to botany and zoology. The Society published periodically large quarto volumes containing information regarding the best methods of horticulture and the introduction of new plants, illustrated with many beautifully colored full-page engravings.[2]

This was the institution under whose auspices Douglas was to accomplish so much for the botanical world in the next few years.

It came about in this way. Hooker was a friend of Secretary Sabine's, and to him he recommended the young man for an appointment as botanical collector. Stewart Murray joined in the recommendation, though probably with some regret, since he had promoted Douglas to the position of head gardener[3] and

[2] *Transactions*, Horticultural Society of London, vol. I-VII (1820-1830); DeWitt Clinton, memorandum, "Horticultural Society of London" (1823), MS., Columbia University Library, New York City. The *Transactions* are now highly prized by libraries and collectors.

[3] W. J. Hooker, "On the Botany of America," *Brewster's Edinburgh Journal of Science*, no. III (1825), p. 108, reprinted in *American Journal of Science and Arts*, New Haven, IX (1825), 263.

would be losing a good man. Accordingly, Douglas was engaged by the Horticultural Society and in the spring of 1823 he left Glasgow for London.

On his arrival he was made familiar with the Society's garden —the new thirty-three acre tract at Chiswick. Although begun only the previous year it already had three thousand fruit trees in its orchard, twelve hundred varieties of roses, an arboretum for ornamental trees and shrubs, and a fruit and kitchen garden for experimental purposes; soon it was to form the most complete collection of fruit trees and hardy trees and shrubs in the world. Here experiments of various kinds were conducted to verify and test former results and to try new practices. Strange plants were received from all parts of the world, sent by the society's collectors and corresponding members, by foreign botanical societies, and by British army and naval officers, travelers, and missionaries. These were propagated in the garden, and the seeds or cuttings were distributed among the society's members. At the same time English plants and seeds were sent abroad.[4]

Collectors had already gone to India, South America, the east and west coasts of Africa, and the West Indies. At first it was proposed to send Douglas to China,[5] but news having come of trouble there between the English and the natives, it was decided that he should go to the United States.[6]

This was not to be a journey to a remote country unknown to botanists; such expeditions were to come later. The botany of the eastern half of North America had become fairly well known in Europe. Although, during the early years of the United States, its own people were too fully occupied with agriculture and commerce to devote much time to the study of their native botany, it was not neglected by Europeans. André Michaux,

[4] In 1904 the gardens were moved to Wisley, near Weybridge.

[5] Chile, according to *Transactions*, Horticultural Society of London, V (1824), iv. Both China and Chile were then in unrest.

[6] W. J. Hooker, "Memoir," *OHQ*, V, 228; Minutes of Council, Horticultural Society of London (1823), Royal Horticultural Society, London.

on behalf of the French government, spent some years in botanizing from Florida up toward Hudson Bay, and in 1803 published his *Flora Boreali-Americana*—all in Latin—the first *Flora* of North America. Frederick Pursh, an Austrian, had the privilege of classifying the plants collected by Lewis and Clark on their overland journey to the Pacific, and published his *Flora Americae Septentrionalis* in 1813—the specific characters in Latin, the observations in English—enumerating about double the number of Michaux's species. Thomas Nuttall, a Yorkshireman, spent many years in America, part of the time as instructor in botany at Harvard University, botanizing principally in the South and West. His *Genera of North American Plants*, Philadelphia, 1817—entirely in English—included all the species of earlier authors as well as many additions, and, according to Hooker, formed "an era in the history of American botany." [7]

Nor was Douglas the first of Hooker's pupils to collect in America. John Goldie had visited the United States and Canada during 1817-1819.[8]

Although the Americans had been too busy to devote much time to the scientific side of botany they were advancing in practical horticulture. Since the introduction of European vegetables and fruits by Jacques Cartier, beginning in 1541, great progress had been made. As early as 1609 an apple had been developed which was reported to Europe as "marvellously big and full of a certain juice very delicate and which intoxicates as much as wine." [9] Not long after, the first orchard in New York was planted. It was part of the *Bouwerie* farm of Governor

[7] W. J. Hooker, "On the Botany of America."

[8] John Goldie, *Diary of a Journey in Upper Canada and some of the New England States, 1819* (Toronto, 1897); plants described in *Edinburgh Philosophical Journal*, VI (1822), 319-332. He afterwards returned to live in America, settling near Ayr, Upper Canada, and founding the well-known Ontario family of that name (John Goldie [a grandson], "In Memory of David Douglas," *British Columbia Historical Quarterly*, Victoria, B. C., vol. II, 1938, pp. 89-90).

[9] *Proceedings and Transactions of the Royal Society of Canada*, Ottawa (1905), sec. II, p. 16.

Peter Stuyvesant, laid out in 1647 on the site of the present Bowery in New York City. From this orchard grafts were disseminated up the Hudson and to inland points. In the period following the American Revolution interest in botanical matters increased—as it did in Europe during the same period. However, there was a different emphasis: in Europe it was on the colorful; in America it was on the useful. Hence, fruit and vegetable culture were favored; and under the leadership of two of the early presidents, Washington and Jefferson, both practical horticulturists, it began to make headway.

The primary object of Douglas's visit was to investigate the latest developments in fruit-growing and to obtain samples of new trees. He was also to collect any interesting plants and seeds, being authorized to travel as far west as Amherstburg in Upper Canada.[10]

[10] Letter, May 30, 1823, Joseph Sabine, Secretary, Horticultural Society of London, to DeWitt Clinton, Governor of New York, Columbia University.

3

UNITED STATES AND UPPER CANADA

DOUGLAS left London by stagecoach early on June 3, 1823, and after traveling all night reached Liverpool next afternoon. On the way he studied the roadside flowers and hedgerows, and during changing of horses got out to botanize. He found a floating heart (*Limnanthemum nymphoides*) he had never before seen in its natural state.

At Liverpool he spent all his spare time examining the "treasures" of the Botanic Garden, many of which had come from North America. Sailing next day on the *Ann Maria,* Captain Tair, they were unable to clear the river owing to lack of wind. This caused "great mortification" to the captain, says Douglas, "but was truly goodness to me"; it enabled him to return for a second visit to the Botanic Garden.

On board again the following morning, the ship was towed out to sea by steam tugs and left to toss aimlessly in a contrary wind for two days. Most of the passengers were seasick. Douglas escaped. A changing wind then filled the sails and the long, tedious voyage began. A week after sailing water was rationed —two quarts to each person per day. During a heavy shower Douglas "could not but observe how the dogs eagerly licked the decks. Some of passengers washing their clothes." As for food, many of those who "found for themselves" ran short and had to replenish their supplies by purchase or barter from one another, "which made good sport." Among the crew tobacco became so scarce that it was used twice over; after being chewed it was dried and smoked.

Douglas found interest in the clouds of seafowl that floated above the ship and the enormous schools of porpoises that surrounded it. He shot some of the former, but lost them in the strong current. While passing the Azores he observed through the captain's glass the small shrubs on one of the islands and was disappointed at not being able to land for a closer examination. He noted the variations of the weather daily in his diary. He studied some botanical books and a Spanish grammar he had brought, the latter in accordance with the wishes of Sabine, who already had seen the possibility of using him on a mission to some Spanish-speaking country.[1]

So the time passed. The fifty-nine-day voyage was marked only by one untoward incident, which occurred the day before landing:

> The ship this morning was all in an uproar, in consequence of a horse, which one of the passengers had, being looked on as dying; it cost him £200 in England, and after troubled passage the poor man lost his horse.

Landing at New York on August 3, and feeling like a prisoner set at liberty,[2] he at once began his work. The wild growth of Staten Island attracted him, "the oaks and maples growing spontaneously," the wild berry bushes and the many wildflowers, including the verbena, smilax, partridge berry, prince's pine, eupatorium, ironweed, and the lowly chickweed. At the Fulton Street vegetable market he was impressed by the large beets, carrots, and onions among the vegetables, and the melons, apples, and pears among the fruits.

Two New Yorkers, Dr. David Hosack and Thomas Hogg, helped greatly by showing him about the city, which even then was no small place, having a population of over 125,000. Hogg, recently come from England, had set up in business as a nursery-

[1] Copy of letter, Oct. 1, 1823, Joseph Sabine to Douglas, Royal Horticultural Society, London.

[2] Douglas, *Journal*, pp. 78, 80.

man and florist at the present corner of Broadway and 23rd Street. Hosack, a distinguished physician and largely responsible for founding Bellevue Hospital, was equally famous in botany for having established the Elgin Botanical Garden, the first of its kind in New York. Failing to enlist public aid, he had founded it in 1801 entirely at his own expense, buying twenty acres for the purpose on what is now Fifth Avenue between 47th and 51st Streets. In 1810 he sold the garden to the State of New York at a loss of $28,000, the state later turning it over to Columbia College. However, owing to lack of financial support and public interest (it was too far out of the city), the garden became neglected, and at the time of his visit Douglas was sorry to find it actually in ruins. Eventually, when the city moved northward, the property was the chief source of income of the College.[3]

Fruits being Douglas's chief concern, he visited some of the fine orchards across the Hudson, including one of an old Dutch farmer who specialized in peaches and plums. Peaches, first used for fattening pigs and later for making brandy, by this time had taken their proper place in the horticultural realm and become prominent in American orchards. After giving due attention to the methods of fruit culture Douglas turned to the old Dutchman's ornamental garden, and spent a whole day going over *Sarracenia purpurea* (side-saddle flower), *Lycopodium* (club moss), *Nymphaea* (yellow pond lily), *Nuphar* (yellow water lily), *Orontium* (golden club), *Chironia,* and other plants.

He was interested to learn the remarkable origin of the famous Washington plum—a name which, he dryly observed, "every product in the United States that is great or good is called"—and was informed by Hosack that it came from a tree grown (without grafting) from a sucker from a tree which had been killed nearly to the roots by lightning.

[3] Ten miles northward is not too far out for the Botanical Garden of today. Where Hosack's plants flowered now stands Rockefeller Center (Radio City); the land that he bought for $4807 yields millions of dollars annually. What a different crop the old garden site now puts forth!

After a few busy days in New York and vicinity Douglas and Hogg went by steamer and stagecoach to Philadelphia. Although not on his intended itinerary, this city, founded by the garden-lover William Penn and possessing the oldest and best botanical gardens in America, could not be ignored. Moreover it had an intimate connection with his native county in the clump of trees at Blair Drummond which had been planted by Benjamin Franklin, the great Philadelphian, while on a visit to Scotland.[4]

Douglas appreciated the nicety of the Philadelphia gardens and the superiority of their plants and trees. Everything was much finer than at New York. This was to be expected, notwithstanding that Philadelphia was only half as big, since it was in keeping with the fundamental differences between the two cities. New York was a bustling, hustling, dirty place, where pigs in the streets were the scavengers,[5] while Philadelphia showed its Quaker origin in the quietness, cleanliness, and piety of its inhabitants. The upper classes were better informed and more refined, and the lower ranks more law-abiding, than in New York. Indeed, it was said that "no two cities within a hundred miles of each other, can differ more than New York and Philadelphia." [6]

He found particular interest in the plant specimens collected by the Lewis and Clark Expedition to the Pacific, 1804-1806, and the Long Expedition to the Rocky Mountains, 1819-1820, which had been distributed among Philadelphia institutions. Some were in a small plot at the University of Pennsylvania, kept up by the professor of botany, William P. C. Barton, and the university janitor, William Dick, jointly at their own expense (the beginning of the present University Garden).

[4] Thomas Hunter, *Woods, Forests, and Estates of Perthshire* (Perth, 1883), p. 254.

[5] Nicholas Garry, "Diary of Nicholas Garry, Deputy-Governor of the Hudson's Bay Company from 1822-1835," John Bourinot, ed., *Proceedings and Transactions of the Royal Society of Canada* (1900), sec. II, p. 83.

[6] John Howison, *Sketches of Upper Canada* (Edinburgh, 2nd ed., 1822), pp. 349, 353.

More Northwest plants were to be seen in the gardens of Bernard McMahon, to whom President Jefferson had entrusted a good portion of the seeds brought back by Lewis and Clark, and at the Woodlands, "the finest American establishment," according to Douglas, and known far and wide for its large and remarkable collection of exotic plants from various parts of the world. (Only the name now remains, Woodland Cemetery being part of the old estate.)

He was also interested in the zoological specimens collected by the western expeditions which were deposited at Peale's Museum,[7] at that time a famous depository for natural history curiosities.

Vegetables and fruits much superior to those in New York were found at the great High Street market, which was about half a mile long, and for quantity and variety was not exceeded by any in the world.

Leaving the Quaker City on August 27 with many plants which had been presented to him, and noting at Bordentown the well-kept fields and pleasure grounds of Joseph Bonaparte, ex-king of Spain and brother of Napoleon, Douglas and Hogg arrived in New York next evening.

By this time considerable interest had been stirred up regarding his visit to America—most of it probably by Dr. Hosack. The *New York Statesman,* on August 26, 1823, published an article two thirds of a column long, telling of the work and itinerary of the "eminent botanist," and concluding with a plea for greater interest by Americans in horticulture, a subject "but little understood in a country which opens a wider field than any other on the globe." A special meeting of a committee of the New York Horticultural Society was called for the purpose of showing its good will to Douglas and the London Society, and this proved of real assistance to him in selecting and obtaining suitable fruit trees and plants to take back to England. At the close of the meeting he was taken home for the

[7] Douglas, *Journal,* p. 246.

night by the president, Martin Hoffman, prominent merchant, whom he summed up as "a very worthy respectable gentleman. His knowledge of gardening is but limited, but he takes a very spirited interest in promoting the science."

The hospitality and courtesies extended to Douglas throughout his tour exemplified the spirit of generosity that was permeating America. An era of expansion and prosperity was in progress. The original thirteen colonies had increased to twenty-four states, two of them west of the Mississippi, and the population of the country had trebled. Civilization was advancing toward the Stony (Rocky) Mountains and apparently would not stop short of the Pacific Ocean. With growth and prosperity came a feeling of good will and neighborliness. Douglas met with it wherever he went. It had even begun to overflow the national boundaries. It was the year of the Monroe Doctrine.

After a few days spent in looking after and cataloguing the plants obtained in Philadelphia and in visiting nurserymen and orchardists, he left on September 4 by Hudson River steamboat for Albany, stopping over for a day to visit lawyer James Thomson's gardens and woodlands at Elerslie (near Clermont). He greatly enjoyed the scenery along the river.

At Albany, the capital of the State of New York, he called on former governor DeWitt Clinton, then engaged with the construction of the great public work for which he is famous, the Erie Canal.[8] Besides being one of the leading statesmen and economists in America, Clinton had many other interests, among them horticulture. He was a member of the Horticultural Society of London, and he and Dr. Hosack were its most active American correspondents. Sabine had written apprising him of Douglas's mission and stating that "whatever protection, advice or assistance you give him will be very gratefully acknowledged by us."[9] Accordingly, Douglas was kindly received

[8] Clinton was reëlected governor the next year. The genus *Clintonia* (beartongue or Clinton's lily) is named after him.

[9] Letter, May 30, 1823, Columbia University.

and given helpful advice in regard to his journey to Upper Canada, with letters of introduction to men of science or influence along the route.

Starting from Albany by stage on the eighth, the bad road and the jolting "reduced" him so much that at Little Falls he changed to canalboat on the Erie Canal, arriving at Rochester on the eleventh. The boat, horsedrawn and traveling about four miles an hour, was "fitted up on good principles; accommodation for twenty-four ladies in one cabin and as many in the men's. I slept on board but was much disturbed passing the locks."

Rochester (just then beginning its commercial development), with its mills, factories and foundries, had no attractions for him. He was interested in the products of nature rather than those of man, so he left quickly by stagecoach and hurried on, stopping overnight at Avon, where he visited the celebrated sulphur springs and saw many health-seekers.

All next day he traveled on the famous old Seneca Turnpike, or Genesee Pike, breakfasting at Caledonia, "a settlement of Scotch people," passing through Batavia, "a neat little village," and reaching Buffalo at midnight. On the thirteenth he embarked on the steamship *Superior,* and after a pleasant passage of sixty hours on Lake Erie, with calls at Dunkirk, Erie, Fairport, Cleveland, and Sandusky Bay, arrived off Amherstburg, Upper Canada. As American ships were not allowed to land passengers at Amherstburg, he was put ashore on an island in the Detroit River, and was taken across to the village by an Indian in a small birch-bark canoe. He was warmly welcomed by Henry Briscoe, a friend of Sabine's brother, Captain Edward Sabine,[10] who had served in Canada during the war of 1812-1815.

The settlements along the Canadian side of the Detroit River were the first in Upper Canada (Ontario), the original settlers,

[10] David Douglas, Journal of part of his eastern American travels, MS., Royal Horticultural Society.

chiefly French-Canadians, having been drawn to that remote region by its agricultural possibilities. Undaunted by the dangers of frontier life they laid out their farms in long narrow strips, which were awkward for farming but were convenient for mutual protection against the Indians. Fruits did especially well in the light, sandy loam, and at the time of Douglas's visit the apples, pears, plums, peaches, and grapes exceeded in size, beauty, and flavor those raised in any other part of the colony, reaching such a high state of perfection that the district was known as the "Eden of Upper Canada." Amherstburg was the chief settlement and the garrison town, with a population of about one thousand.

The day after his arrival Douglas, accompanied by Briscoe with his dogs and gun, went on a combined hunting and botanical trip east of the town. This was his first real outing in the American forest, and, although Briscoe was "a marksman of the first sort," hunting seems to have given way to botany. The astonishing size of the oak and walnut trees, the fine crops of Indian corn, the fields of tobacco cultivated for the far-off Montreal market and prophesied by Douglas to become soon "an article of great importance to our Canadas," and the asters, sunflowers, goldenrod, ironweed, and gentian, all yielding him seeds, so pleased Douglas that he declared: "This is what I might term my first day in America."

Next morning, unaccompanied by Briscoe or the dogs, but taking the gun, he went to Bois Blanc Island, opposite Amherstburg. Here he spent the day roaming among the large oaks after which the island is named, and among a profusion of hawthorn and honeysuckle. He used the gun to shoot down branches, leaves, and acorns from the oaks for his botanical collection. A honeysuckle which he discovered was received in England as a great novelty (*Lonicera hirsuta*).[11]

The next few days were occupied in a journey to Lake St.

[11] Letter, July 26, 1825, Joseph Sabine to Douglas, copy at Royal Horticultural Society; *Transactions*, Horticultural Society of London, VII (1830), 244.

Clair and the River Thames to collect plants, which later were to bloom in gardens along the more famous Thames. He rode as far as Sandwich with Dr. Robert Richardson, physician to the Indian Department and judge of the District Court, admiring en route the well-cultivated fields and neat gardens of the French settlers. Continuing to Lake St. Clair with a hired horsedrawn "car" and man, he had an annoying experience. He climbed an oak tree in order to obtain seeds and specimens:

The day being warm, I was induced to take off my coat and in that state I ascended. I had not been above five minutes up, when to my surprise the man whom I hired as guide and assistant took up my coat and made off as fast as he could run with it. I descended almost headlong and followed, but before I could make near him he escaped in the wood. I had in my pockets my notes and some receipt of money, nineteen dollars in paper, a copy of Persoon's 'Synopsis Plantarum,' with my small vasculum. I was thus left five miles from where I had left the car, in a miserable condition, and as there was no remedy that could be taken to better myself, I tied my seeds in my neckcloth and made to my lodging. I had to hire a man to take me back to Sandwich as I could not drive; and the horse only understanding the French language, and I could not talk to him in his tongue, placed me in an awkward situation. I had to borrow a coat as there was no tailor to make me one. On my getting to Sandwich I remonstrated with the man who recommended my assistant to me, but he said that he never did so to his knowledge, and so on. However, I found my guide was a runaway Virginian.[12]

Returning to Amherstburg, he crossed the river to Michigan for a day's botanizing. Detroit he did not visit; it was only a frontier town of about 1500 people.

He thought Amherstburg and vicinity "certainly a fine field and would afford an abundant harvest," but rainy weather, the loss of his money, and the removal of his friend Briscoe to Kingston, combined to cut short his stay; and on September 25

[12] Slavery had long been outlawed in Upper Canada (the first jurisdiction in the world to take that step), and many fugitive slaves from the United States, traveling by the "Underground Railroad," had found refuge in the Amherstburg district.

he left by steamboat for Buffalo. The passage was very rough and he "experienced a motion that could not be surpassed in an ocean," the ship arriving at Buffalo with one of its wheels swept away.

A three-day side trip to Niagara Falls followed. He was "sensitively impressed with their grandeur," but gave most of his attention to the surrounding vegetation—plants on the sides of the cliffs, the trees and flowers along the banks of the river and on Goat Island—and collected several specimens.

At Lockport he visited David Thomas, chief engineer of the Erie Canal, who was also the pioneer nurseryman in western New York and had in his orchards at Aurora the most extensive and valuable collection of fruit trees west of the Hudson. They spent a morning together botanizing in the beech woods north of Lockport. Proceeding on to Rochester he traveled over fifteen miles of corduroy road through a swamp "which rendered it very unpleasant."

October 8 was a gala day at the staid old Dutch city of Albany; the opening of the eastern section of the Erie Canal was being celebrated. Douglas found "the town was all in an uproar —firing of guns, music, &c.," but instead of joining in the fun he quietly turned to his seeds, unpacking and arranging them and putting them in fresh paper. Two or three days were spent in botanizing and in calls on ex-Governor Clinton and some of his friends: Mr. Tracey, who had an extensive herbarium; Stephen Van Rensselaer, wealthiest man in the United States and a leading horticulturist; and Jesse Buel, retired journalist and printer, who had taken an abandoned tract of sand and made it one of the best farms in the state. Clinton found time from his official duties to entrust Douglas with a case of six wild pigeons and boxes of minerals, apples, pears, moccasins, books, and even speeches, all for his friend Sabine of the Horticultural Society in London.[13] On the Van Rensselaer estate Douglas noted "a large space of ground occupied as pleasure or flower garden,

[13] Draft letter, Oct. 10, 1823, Clinton to Joseph Sabine, Columbia University.

which is a novelty in America, as little attention is paid to anything but what brings money or luxury for the table. . . . Mr. van Ransaleer is a man of taste."

Although it was his duty to visit as many of the fine orchards and gardens as possible, what he enjoyed most was to botanize out in the wilds, and for this he took advantage of every opportunity. Here at Albany he spent practically a whole day in hunting for the curious *Pterospora andromedea* (pine drops), finally finding it in a small ravine two miles south of the town. Next day a twelve-mile walk was taken to look for some *Erythraea,* but this time his efforts were in vain owing to recent rains' having inundated the ground.

Leaving Albany on October 11, Douglas stopped at the home of General Morgan Lewis at Clermont. Here he was taken ill with rheumatism in his knees, and was so lame that he was compelled to remain for six days. The illness probably was caused by traveling in cold stagecoaches and canalboats, especially at night, and by exposure during his botanical tramps. More pigeons were obtained from the General's neighbor, James Thomson, together with apple trees and other plants. Returning to New York on the 19th on the steamboat *Chancellor Livingston,* which was "celebrated for magnitude and elegance," his time was taken with visits to nurserymen to arrange for trees and plants to take back to England. Among them were Grant Thorburn, first seed merchant in New York and founder of seed farming in America, William Wilson, pioneer market gardener, and Michael Floy, prominent horticulturist.

He attended a meeting of the New York Horticultural Society, then in a flourishing condition and numbering many prominent citizens on its membership roll, and he was "glad to see it in such a state of perfection," a state, however, which did not last, for by 1844 the organization was defunct.[14]

[14] Reorganized about 1852, it became defunct again by 1889. The present organization, the Horticultural Society of New York, was formed in 1900.

Instead of returning at once to England, Douglas was persuaded by Clinton and Hosack to postpone his departure for a few weeks in order to extend his collections.[15] So another journey was made to Philadelphia. At the University Garden, janitor-gardener Dick's seed cabinet was generously placed at his disposal. He also had the good fortune to meet Thomas Nuttall, who had collected on the Missouri River and in the Southwest, and he obtained useful information regarding new plants from those regions. Nuttall had high yet simple ideals of life—ideals which no doubt appealed to young Douglas. For nearly ten years he had traveled through America "with no other end in view," he says, "but personal gratification; and in this I have not been deceived; for innocent amusement can never leave room for regret. To converse, as it were with nature, to admire the wisdom and beauty of creation, has ever been, and I hope ever will be, to me a favourite pursuit; and to communicate to others a portion of the same amusement and gratification has been the only object of my botanical publications." [16]

Nuttall and Douglas visited the famous Bartram garden at Kingsessing on the Schuylkill River. This, the oldest botanical garden in America, dating from 1731, and now part of Philadelphia's park system, owes its origin to John Bartram, a poor ploughman whom a single daisy inspired to become a great botanist. Resting one day from his plough, he casually plucked the flower and examined it. His interest was aroused and he determined to know more about daisies, and other flowers as well. He studied Latin in order to read books on botany; he dissected the flowers and weeds on his farm; then, buying a piece of waste swamp land at a tax sale, he traveled the country from Florida to Canada to collect specimens for his "Garden of Delight," which became famous in both Europe and America. He was finally regarded (according to no less an authority than Linnaeus) as "the greatest living botanist in the world."

[15] Draft letter, Oct. 28, 1823, Clinton to Joseph Sabine, Columbia University.

[16] Thomas Nuttall, *A Journal of Travels into the Arkansa Territory during the Year 1819* (Philadelphia, 1821), preface.

Douglas saw the great cypress tree grown from a riding whip planted by Bartram eighty-five years before. The tree is gone now but the trunk still attracts visitors. A sorrel tree also drew his attention. Bartram's son, who had died not long before Douglas's visit, "tried for upwards of forty years to raise it from seed, and always was unsuccessful till this season, when he had the gratification of transplanting an abundance of them in small boxes two days before he died." Douglas got several of the plants to take to England.

At the well-stocked nursery of David Landreth he was generously treated with a good selection of trees, roses, and fruits; and he noted appreciatively: "He is certainly an attentive and punctual man."

Leaving Philadelphia on November 4, he visited at Burlington, New Jersey, the large orchards of William Coxe, then regarded as the leading authority on fruit-growing in the United States. Here he got samples of several varieties of fruit trees and various seeds and plants, together with two bottles of seven-year-old cider, presents from Coxe to the president and secretary of the Horticultural Society.

Douglas's travels in the United States and Upper Canada were now over. From the names of the persons and places visited it is apparent that he covered the ground thoroughly. He interviewed most of the leading horticulturists, inspected their gardens, orchards, and nurseries, investigated their methods, heard their ideas, studied their plans, and discussed their problems, not forgetting, of course, to collect all the interesting specimens he could. The expedition had served the purpose intended by the Horticultural Society. He had obtained a mass of information and a wealth of specimens.

For him the expedition served a different purpose. It extended his education. It was a postgraduate course, so to speak, in field study of American horticulture, with its specialists as his instructors. His training was still going on.

At New York he had the task of gathering together all the

plants, trees, seeds, and specimens that he had bought or collected during his tour and of packing them up carefully to withstand the long ocean voyage. Many of the fruit trees he dug up himself. The pigeons and some wood ducks and quail had to be attended to. His botanical notes had to be written up. Many calls and farewells had to be made. And of course he could not resist a last opportunity to glean new plants. A two days' expedition to a big swamp across the Hudson was made with Hogg to procure specimens of the pitcher plant (*Sarracenia purpurea*) and other bog herbs. After a good deal of wallowing about in the mud and muck, in which Hogg went down "to the middle," they succeeded in reaching the plants, carried them two miles through the swamp, and returned to town besmeared with filth but happy.

The Horticultural Society's collector left America with kind feelings toward all he had met, with one exception. He had been treated hospitably everywhere save at the Prince nursery —the Linnean Botanic Garden of William Prince and Son at Flushing—reputed to be the leading nursery in New York. It specialized in fruits, particularly plums. (Later on it got into difficulties and eventually was foreclosed.) Douglas and the Princes had poor opinions of one another from the start. The keenly discerning Scot was disappointed to find that, notwithstanding the nursery's great reputation and extensive catalogue, Prince was "a man of but moderate liberality" and "most of his ground is covered over with weeds." Moreover, he thought Prince avaricious for charging two hundred dollars for his plants and demanding free copies of the Horticultural Society's *Transactions*, so he obtained plants elsewhere as far as he could. This so incensed Prince that he got his son to write secretary Sabine a scurrilous letter regarding Douglas.[17] Douglas had acted in accordance with his instructions, however, which were to obtain plants as presents whenever he could, otherwise to buy them where they could be obtained cheapest and best, being

[17] Letter, Oct. 3, 1825, Douglas to Clinton, Columbia University.

particularly careful to have them true, the propriety of each case being left to his judgment.[18]

He had the satisfaction of getting one plant from Prince which he took home as a great ornamental novelty, the tall Oregon grape (*Berberis aquifolium*). It had been introduced from the Far West by Lewis and Clark.

Douglas was particularly grateful for the kind assistance given by Hosack and Clinton, and found it impossible to speak of one more highly than the other: "I can say nothing more than that Dr. Hosack is a Clinton, and Clinton a Hosack."

Leaving New York on December 10, he reached London on January 9 (only about half the time required for the outward journey). The ship was the *Nimrod.* The only noteworthy events of the voyage were two: first, a severe attack of seasickness, not on Douglas, but, most remarkable, on the ducks, which "were very sick for two days and ate nothing"; and, second, a bloody engagement between Clinton's pigeons and Thomson's, which Douglas "most stupidly . . . put in the same cage . . . They fought furiously and 4 fell in the engagement 2 on each side."[19]

[18] Letters, Oct. 1 and 5, 1823, Joseph Sabine to Douglas, copies at Royal Horticultural Society.

[19] Letter, Oct. 3, 1825, Douglas to Clinton, Columbia University.

4

FIRST GREAT ADVENTURE

THE JOURNEY to eastern America had been successful. His employers were well satisfied. "This mission was executed by Mr. DOUGLAS with a success beyond expectation," said the Horticultural Society's official publication. "He obtained many plants which were much wanted, and greatly increased our collection of fruit trees by the acquisition of several sorts only known to us by name. It would be unjust here to omit mentioning the uniform kindness and attention with which he was received in every part of the United States that he visited. It is most gratifying to have to add, that the PRESENTS of cultivated plants to the Society embraced nearly everything which it was desirous to obtain; and that the liberality with which they were given was only equalled by the hospitality with which the Collector was received." [1]

The trees and plants obtained from Michael Floy were especially featured, several pages being devoted to them. Detailed descriptions were given of the twenty-one varieties of peaches and ten varieties of apples, and of the cuttings from Governor Stuyvesant's Pear, a tree over 170 years old which had been imported from Holland.[2]

Naturally, Douglas had the job of setting out the plants and seeds he had brought home and of looking after them. All arrived in good condition. Indeed, some of the beautiful *Pyrola* (wintergreen) and *Epigea repens* (trailing arbutus or mayflower) were already in blossom owing to the warmth of the

[1] *Transactions*, V (1824), v.
[2] *Ibid.*, VI (1826), 409-418.

Gulf Stream.[3] Among the new or rare ones which flowered in their new home were *Oenothera speciosa* (a perennial border plant), *Oenothera triloba* (a stemless spreading annual), *Pogonia pendula, Triphora pendula,* and *Pedicularis canadensis* (a lousewort).[4] Success also attended the growing of his seeds of ice lettuce, which was said to belong to "the division of Silesian or Batavian Lettuces, and must not be confounded with the Ice Lettuce of Scotland, which is our White Cos Lettuce." [5]

The pigeons drew favorable comment. Douglas was proud of "the complements lavished on them and even by Ornithologists of merit. Its fine plumage, its chaste and dignified attitude. A pair were sent to the Earl of Liverpool, another to Joshua Brooks, Esq.[6] for dissection who was pleased with its proportions. The former valued them beyond measure." One of the wood ducks died, but the survivor "got a British consort. They live mutually, and reign paramours of a spacious pond in the ornamental department of the Society's garden." [7]

Favorable opinions of his mission also reached the Society from America. Governor Clinton wrote that he considered their appointment of Douglas "a judicious one. He unites enthusiasm, intelligence & persevering activity." [8] Again, "He appears to me to combine the essential qualities required in trusts of this nature." [9] The New York Horticultural Society showed its esteem by making him a corresponding member.[10] Indeed, he had done his work so thoroughly that when Dr. John Richardson visited New York two years later he found some of the botanists there "furious against Douglas for having

[3] Letter, Oct. 3, 1825, Douglas to Clinton, Columbia University.

[4] *Transactions,* Horticultural Society of London, VI (1826), 293-296.

[5] *Ibid.,* p. 575; Douglas, *Journal,* p. 337.

[6] Joshua Brookes (1761-1833), famous English anatomist, dissector, and teacher.

[7] Letter, Oct. 3, 1825, Douglas to Clinton, Columbia University.

[8] Draft letter, Oct. 10, 1823, Clinton to Joseph Sabine, Columbia University.

[9] Draft letter, Oct. 28, 1823, Clinton to Joseph Sabine, Columbia University.

[10] *Constitution and Bylaws,* New York Horticultural Society (New York, 1824).

deprived the Flora of New York of several rare plants which he carried off root and branch from the only localities where they are known to exist."[11] Perhaps the furious ones consisted of Prince and Son.

Besides his garden duties, the arranging of his dry plants, and the transcribing of his notes, Douglas undertook to put in order the herbaria received by the Society from voyagers into the Northern and Arctic regions, and made up a collection to send to Dr. Hosack for an American institution.[12] Considerable time also was spent in writing a lengthy account of the American oaks. Of the thirty-four species enumerated by Pursh, Douglas had met with no less than nineteen, and the account contained his personal observations on these, together with observations taken from Michaux's work. Additions were made after his return from northwest America some years later. Although evidently written for publication by the Horticultural Society (to whom it was sent accompanied by a letter to the secretary), it was not published until 1914, when it was included with his journals.[13]

Notwithstanding his travels in the United States and Canada and his intercourse with prominent people there, this period of Douglas's life was marked by excessive modesty and shyness. The president of the Horticultural Society found him "the shyest being almost that I ever saw; and upon my requesting . . . to ask him some questions respecting a part of America through which he had travelled, Mr. Sabine said 'now Douglas will be terribly frightened;' and so . . . he appeared to be, until I had talked to him for some time in a friendly and familiar way."[14]

[11] Letter, April 22, 1825, Richardson to W. J. Hooker, Kew Gardens. "You will not find things at New York as comfortable as when you were there" (letter, Feb. 20, 1826, Joseph Sabine to Douglas, copy at Royal Horticultural Society).

[12] Letter, Oct. 3, 1825, Douglas to Clinton, Columbia University.

[13] Douglas, *Journal*, pp. 31-49.

[14] Thomas Andrew Knight, *A Selection from the Physiological and Horticultural Papers . . . by the late Thomas Andrew Knight, Esq., . . . to which is Prefixed a Sketch of his Life*, Frances Acton, ed. (London, 1841), pp. 39-40.

Nevertheless, after a few months opportunity came for another, and a greater, mission. Again America was calling; not, however, the America of Clinton and Hosack, or Franklin and Washington, that he had just visited, but the America of the Far West; it was the America dreamed of by Jefferson; the North Pacific, explored by Captain James Cook (1778), charted and named by Captain George Vancouver (1792-1794), and first reached overland by Alexander Mackenzie (1793); the territory of the mighty Columbia River, explored by Lewis and Clark (1804-1806), and combed by David Thompson and the North West Company's traders and by John Jacob Astor's fur hunters.

While some of the early explorers gave little or no attention to botanical riches, others took pains to discover them. Cook had with him both a plant collector, David Nelson, and a surgeon-naturalist, William Anderson—the first botanists to visit the Northwest Coast. With Captain Colnett (1786-1790), and later with Vancouver, came Archibald Menzies, surgeon-naturalist. On both voyages Menzies collected plants and cuttings, which he took home for English herbaria. Among his discoveries were the salal, the large-leafed maple, the arbutus, the spruce named after him, and the fir later named after Douglas. The conditions under which he worked made it difficult to take home living plants and seeds, so his collections were restricted in the main to dried specimens.[15]

Of the overland journeys, Mackenzie's was a redoubtable and historic dash to the Pacific and back in a remarkably short period of less than four months, and there was time for nothing but travel. The more leisurely expedition of Lewis and Clark gave opportunity for plant collecting, which, in fact, was one of its objects, and many specimens were obtained for the Philadelphia gardens. A considerable number of those collected, however, were lost on the way.

The plants collected by Menzies and by Lewis and Clark

[15] *Menzies' Journal of Vancouver's Voyage, April to October, 1792*, C. F. Newcombe, ed. (Victoria, B. C., 1923).

whetted the appetites of the English botanists and garden lovers; they wanted more. The Horticultural Society obligingly decided to get them; and, in view of the satisfaction he had given on his recent mission, Douglas was entrusted with the task.[16] No doubt Hooker had a hand in the appointment, as he had had in the former one. He was impressed with the "vast extent of highly interesting country" in British Northwest America and the scarcity of botanical knowledge regarding it. He thought a book on the subject was needed to incorporate the botanical discoveries of the Arctic explorers and those of Douglas and others[17]—a work which he undertook, but did not complete until 1840.

Douglas's second mission for the Horticultural Society was "undertaken under the protection of the Hudson's Bay Company."[18] That, of course, was essential, for since its coalition with the North West Company in 1821 its posts were the only civilized (or partially civilized) spots in the immense territory stretching from California to the Arctic and from Red River to Alaska. Douglas was to be afforded the facilities of these posts and given a passage out from England in the Company's annual ship.[19] This was in accord with the Company's policy of encouraging scientific investigation.

The Company's promise of assistance was accompanied by the brief warning that "he will however find the fare of the country rather coarse and be subject to some privations,"[20] a warning which proved to be no exaggeration.

The offer of a passage on the Company's ship *William and Ann*, Captain Henry Hanwell, was received in the spring of

[16] *Transactions*, Horticultural Society of London, V, (1824), v.

[17] W. J. Hooker, "On the Botany of America." The book he wrote to fill this need was his *Flora Boreali-Americana; or, the Botany of the Northern Parts of British America*, originally issued in twelve parts, the first in 1829. It was published in a two-volume edition in 1840 under the authority of the Secretary of State for Colonial Affairs, London.

[18] *Transactions*, VI (1826), iv.

[19] *Ibid.*, V (1824), vi.

[20] Letter, June 24, 1824, Hudson's Bay Co. to Horticultural Society of London, Hudson's Bay Co. Archives, London.

1824,[21] but it was July 26 before she sailed from Gravesend on her long voyage. The interim was occupied by Douglas in making preparations for the expedition and in familiarizing himself with what was known of the botany, zoology, and geology of northwest America. Considerable time was given to examining the collections of Menzies, who was then living in London, and getting his advice on various points. Dr. John Richardson also had information for him about natural history specimens collected on Captain John Franklin's first Arctic expedition and about plans for a second expedition, soon to start.[22]

Douglas's brother accompanied him from London to the ship and bade him a tearful farewell. Tears, however, were soon forgotten in the joy of finding that a Glasgow classmate, John Scouler, was to make the journey too, having been appointed surgeon-naturalist to the *William and Ann*, upon Hooker's recommendation.[23]

As the ship cleared Land's End on a beautiful summer evening a few days later, Douglas had a melancholy turn: "I stood on deck looking on the rocky shores of Cornwall, burnished with the splendour of a setting sun—a noble scene. By degrees the goddess of night threw her veil over it, and my delightful view of happy England closed—probably closed for ever!"[24] He knew the dangers of his undertaking and the risk he was assuming. Nor was there anything reassuring in the record of the three collectors last sent out by the Horticultural Society: one sent to Africa died there; one sent to Bengal and China became ill and died after his return; only one had survived.[25]

Great enthusiasm and determination must have animated him as he embarked on this, his first great adventure. He was

[21] *Transactions*, Horticultural Society of London, V (1824), v.
[22] Letter, Oct. 3, 1825, Douglas to Clinton, Columbia University.
[23] Hooker, "On the Botany of America."
[24] Letter, Douglas to W. B. Booth, *Gardener's Magazine*, London, XI (1835), 271.
[25] *Transactions*, Horticultural Society of London, V (1824), iii.

going to a distant region inhabited only by savages and a few fur traders, where obstacles and hardships abounded and scientific pursuits were little appreciated; yet, difficult as the prospect appeared, he faced it resolutely, even at the risk of never returning. Indeed, his enthusiasm for the discovery of botanical treasures was such that it became a veritable passion, while his determination was so strong that fear had little place with him. His references to religion show that he was fortified by a steadfast faith in a Supreme Power—a faith which must have helped him in the many dangers and hardships through which he was destined to pass. He also had a considerable measure of the saving grace of humor, invaluable in an emergency.

The first stop on the voyage was at Madeira, on August 10 and 11. On landing he hastened to the vegetable market and carefully noted the peculiarities of the various tropical fruits and vegetables. Visits to vineyards followed. The rest of the time was taken by trips with Scouler to the garden of Henry Veitch, London nurseryman, and to a nearby mountain to study the trees and flowers and collect several interesting plant specimens. He left the island "amazingly gratified."

The six weeks' voyage across the Atlantic to South America was not without its scientific opportunities. The slow-moving vessel permitted him to catch specimens of jellyfish, which were duly catalogued; and the Madeira plants gave "fine amusement during some of the tedious hours; put them in fresh paper." He was also interested in the shoals of flying fish, the Mother Carey's chickens and other strange birds, and tried to catch two sea turtles, one of which he estimated to weigh from 250 to 300 pounds.

Rio de Janeiro was reached on September 28. During the stay of two weeks he was the guest of John Dickson, British naval surgeon, a friend of Sabine's, and was treated "in the most handsome manner," Dickson's servants being placed at his disposal to aid in his botanical pursuits. As usual he carefully explored the public market, finding oranges, lemons, and coconuts in great abundance and very cheap, the oranges and lemons selling at one hundred for a dollar. At Botofago he visited the nursery of William Harrison, who had introduced many interesting plants into England. With him and his brother Henry, Douglas made a short trip to Tijuca, admiring the rich luxuriance of the forest, and being "particularly delighted with the varied and endless forms of *Orchideae.*" He notes that William Harrison gave him a letter of introduction to "Messrs. McCulloch, & c., of St Barbara, New Albion, lest we should put in there or visit it at any future period . . . To have friends in such a remote spot of the globe is of great consequence."[26] Later he was to become intimately acquainted with Santa Barbara during an extended visit to California.

Douglas and Scouler were enraptured with the rich vegetation of the tropical country, and, although incessant rains greatly hampered their movements and the uncertain length of stay of the ship prevented any long journeys into the interior, both collected enthusiastically. Scouler returned to town one day making a strange appearance: he had a large vasculum loaded with plants on his back, his pockets were filled with granite specimens, his hat outside and inside was pinned full of insects, both arms were full of plants, and somehow he carried a snake which was reckoned one of the most venomous species to be found.[27] Douglas collected two boxes of plants, which were sent to the Society in London and were received with great pleasure, many being new, and all excellent;[28] and he took with him on the ship nearly two hundred species of dried plants, some of which, not being fully dry, had to be further dried on board.[29] Among the many rare orchideous plants and bulbs

[26] Douglas, *Journal,* p. 89.

[27] "Dr. John Scouler's Journal of a Voyage to N. W. America," *Oregon Historical Quarterly,* VI (1905), 61.

[28] Letters July 26, 1825 and Feb. 20, 1826, Joseph Sabine to Douglas, copies at Royal Horticultural Society.

[29] Plants described by John Lindley, *Transactions,* Horticultural Society of London, VII (1830), 62-64, 67, 70-71.

which he sent to London was a new species of *Gesneria,* which Sabine named *Gesneria Douglasii* in honor of its discoverer.[30]

They left charming Rio with regret on October 14, hoping to return for a longer stay. During the journey down the coast to Cape Horn they passed immense quantities of seaweed. Some of it was caught and measured sixty feet long, the roots proving to be a zoological treasure, for they contained what Scouler called "a menagerie of marine animals." [31] In the air, bleating like lambs, large numbers of albatross surrounded the ship, and the enthusiastic young naturalists caught several with hooks baited with fat pork. One of them measured over twelve feet from tip to tip of the wings and four feet from beak to tail. Seagulls and petrels also were caught in the same way. All were carefully examined and their peculiarities noted, and a few were preserved.

Ten days of stormy, piercing-cold weather and sleepless nights were experienced in rounding Cape Horn. Yet even this bleak, inhospitable region Douglas thought might have botanical interest:

> As to the geography of plants it would be an interesting point to compare those found on the South extremities of the Southern continent with the very extensive & beautiful Flora of the northern in similar Latitudes. Gladly would I spend a few days on the shores of Terra Del Fuego—many novelties are there no doubt, that for the present must be laid aside.[32]

Off the coast of Chile a short stay was made at Juan Fernandez, the island made famous by Defoe's *Robinson Crusoe*. Great was the surprise on finding it inhabited by five Spaniards and a Londoner, William Clark, who were shooting seals, goats, and wild cattle for their skins. Clark's clothing, shelter, and cooking utensils were so primitive and meagre as to excite the pity of the visitors and bring presents from the ship. Douglas noticed

[30] *Gardener's Magazine,* XI (1835), 271.
[31] "Scouler's Journal," pp. 61-62.
[32] Letter, Oct. 3, 1825, Douglas to Clinton, Columbia University.

a nicely bound copy of *Robinson Crusoe* among Clark's few books, but thought Clark "himself was the latest and most complete edition."

Douglas found Juan Fernandez enchantingly beautiful and termed it "the Madeira of the South." During the short stay of two days he secured seventy distinct and exceedingly interesting plants, some of them being from Cruz Bay, where Clark said Crusoe had resided, and where the sailors of the *William and Ann* identified Crusoe's cavern.[33] By way of reciprocity Douglas planted some vegetable and fruit seeds on the island and left others for Clark to plant.[34] On their departure the visitors were presented by Clark with "a fine female goat, but not one of Robinson Crusoe's, for it was young." A week later it was killed and served for Christmas dinner.

The next stop was at James Island, one of the Galapagos group. Here, during three visits of only two hours each, Douglas collected 175 plant specimens and forty-five birds of nineteen varieties. The birds, very numerous and pretty, were so strange to man that they were readily killed with a stick. Unfortunately nearly all the birds he obtained and most of the plants had to be thrown away, since the incessant rain for twelve days after leaving the island prevented their being exposed on deck, and there was no room for them below. He felt the loss keenly: "Never in my life was I so mortified, touching at a place where everything, indeed the most trifling particle, becomes of interest in England, and to have such a miserable collection to show I have been there." [35]

No stop was made along the Mexican and Californian coasts. On February 12 they came to the latitude of the Columbia River, but an approach to the shore was impossible. The weather was so boisterous that they were obliged to lay to day

[33] "Scouler's Journal," p. 65.

[34] According to W. B. Booth, Douglas said he planted the seeds to "add to the comfort of a second edition of Robinson Crusoe, should one appear" (*Gardener's Magazine*, vol. XI, 1835, p. 272).

[35] Douglas, *Journal*, p. 100.

after day for six long weeks. Furious gales threatened the ship, more of them than had been experienced during the whole of the voyage so far, and "a thousand times worse," said Douglas, "than those of the noted Cape Horn." The sandbar at the mouth of the Columbia was then the most dangerous place for navigation on the entire Pacific coast.[36] It was the scene of many wrecks, the *William and Ann* itself being among them four years later, when not a soul lived to tell the story.

After ineffectual attempts to cross the bar during short periods of favorable weather, success was achieved on April 7, 1825, when, with Douglas and Scouler sounding the depths, the vessel entered the river and anchored in Baker's Bay. The long, tedious voyage of eight months and fourteen days was over. To enjoy the sight of land and be free from the motion and noise of the ship was to Douglas "truly a luxury."

As they rounded the point he spied three kinds of trees: a hemlock, a balsam fir, and "a species which may prove to be *P. taxifolia*." This was his first sight of the giant fir which was to become a familiar friend in his travels along the Columbia, and was subsequently to become known as the "Douglas fir." It had previously been reported by the surgeon-naturalist Menzies, who took twig specimens back to England in 1795. From Menzies's account and specimens it had been illustrated and described by Aylmer Bourke Lambert in 1803 in his great work, *A Description of the Genus Pinus*, in which it had been given the name *Pinus taxifolia*. It remained, however, for Douglas to collect cone specimens and seeds, from which the tree was first grown in England.

Landing next day on Cape Disappointment, the first plant that he and Scouler found was the now well-known salal (*Gaultheria Shallon*). Both were delighted. So attracted was Douglas that he "could scarcely see anything but it." The

[36] Colonel Lawrence Kip, U. S. A., "The Indian Council at Walla Walla, May and June, 1855," in *Sources of the History of Oregon* (Eugene, Ore., 1897) vol. I, part 2, p. 4.

salmonberry (*Rubus spectabilis*) also was abundant. He was in his element. The anticipation held throughout the long voyage of "ranging through the long wished-for spot" and of resuming his "wonted pursuits and enjoyments" was being realized. A virgin territory awaited him; never before had the primeval forests of the Columbia felt a botanist's footsteps.

Also of interest were the natives who visited the ship: their flattened foreheads; their nose and ear ornaments; the skin robes of the men, pronounced by Scouler to be "alike useless for the purpose of decency or of comfort," and the more modest petticoats of the women;[37] their peaceful disposition toward the whites coupled with distrust among themselves;[38] and their dexterous bartering, by means of signs and a few words of English, of hats and other homemade articles for knives, buttons, and other trinkets.[39] On shore a few days later Douglas and Scouler saw—and heard—an Indian war dance by fifty warriors clad in arrow-proof elk skins and war paint, their antics being accompanied by howling, yelling, and the rattling of shells, making all together a grotesque pageant.[40]

They landed from the *William and Ann* at Fort George (Astoria) on April 12. Captain Hanwell had shown them much kindness during the voyage. Probably not many sea captains of that day would have given such attention to scientific passengers as Douglas and Scouler received, nor have allowed them to turn the ship into a floating dissecting room. For his kindness, and for the care given to the seeds and plants sent home by Douglas later, the Captain was presented with the Horticultural Society's large silver medal.[41] No doubt this was done upon Douglas's recommendation.

37 "Scouler's Journal," p. 163.
38 *Ibid.*, p. 165.
39 *Ibid.*
40 *Ibid.*, p. 167.
41 Presented between May 1, 1826, and May 1, 1827 (*Transactions*, vol. VII, 1830, p. 2).

5

COLUMBIA RIVER REGION

DOUGLAS arrived on the Columbia while the headquarters of the Hudson's Bay Company were being moved from Fort George, on the south shore of the river near its mouth, to a new establishment (christened "Fort Vancouver") on the north shore about a hundred miles up. The change was made because of the unsettled international boundary and in the hope that eventually the north shore would be British territory.

The Company's regional chief factor, or district superintendent, was Dr. John McLoughlin, who had entered the fur trade twenty-two years earlier as surgeon to the North West Company, but was now just beginning his long and notable career in the Oregon country. Under him, in charge of the more important outposts, were chief traders, who, like the chief factors, shared in the Company's profits, though to a lesser extent. Clerks or postmasters had charge of the minor posts. Also included in the "gentlemen," or officer class, of the Company were the apprentice clerks—many of them Scottish—from whom were developed the clerks, chief traders, and chief factors. The employees—interpreters, boatmen, and laborers—were mostly French Canadians, Indians, and halfbreeds; among them were also a few Kanakas (Sandwich Islanders). Such were the men with whom Douglas was to associate during the next two years. Above them in the Company were the governor and council (of chief factors, meeting annually at York Factory or Norway House), who had administrative control of the Company's affairs, and the London governor and committee (directors), who had financial and ultimate control.

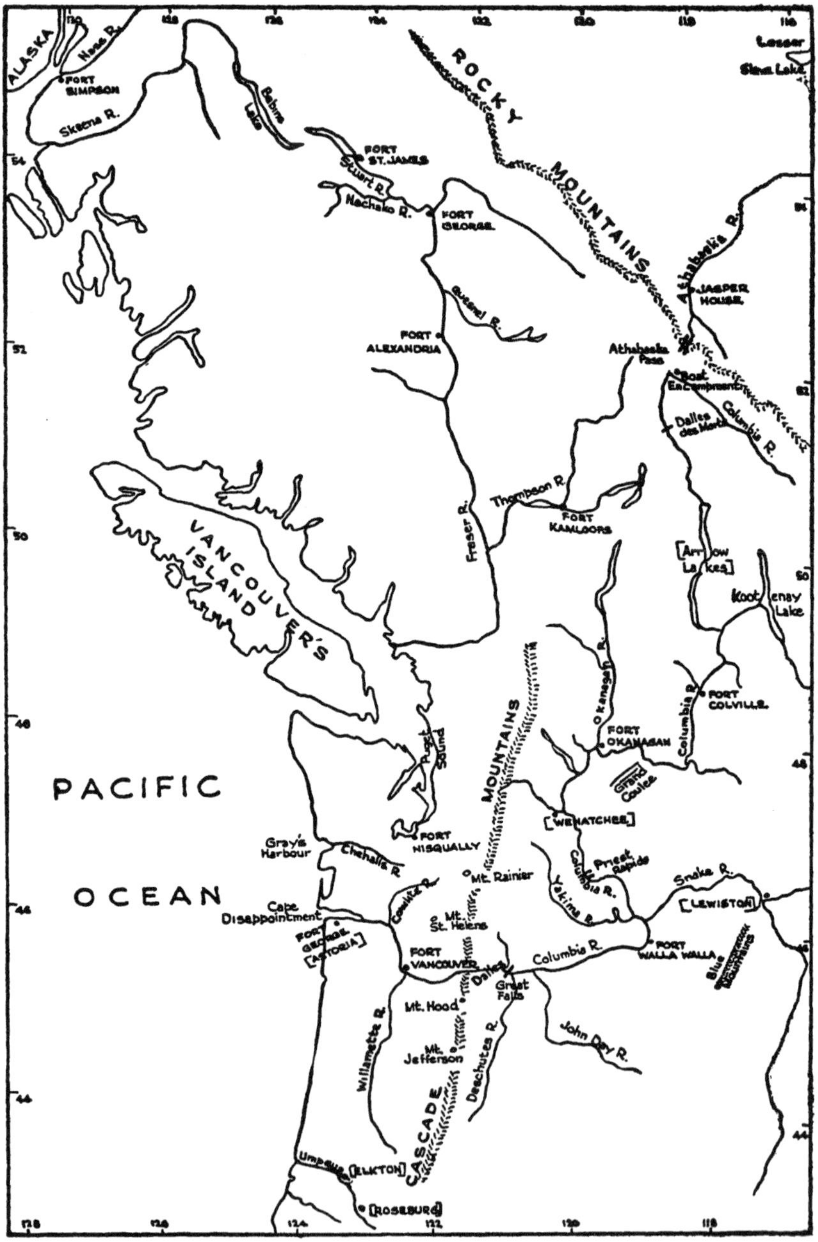

PART OF THE PACIFIC NORTHWEST *as Douglas found it*

(Scale: 1 inch = 140 miles)

Chief Factor McLoughlin came down from the new post to meet the ship and, in Douglas's words, he "received me with much kindness. I showed him my instructions and informed him verbally the object of my voyage, and talked over my pursuit. In the most frank and handsome manner he assured me that everything in his power would be done to promote the views of the Society." [1]

When McLoughlin left Fort George on April 19 to return to Fort Vancouver, Douglas went with him, traveling in a canoe manned by a Canadian and six Indians. After going forty miles they landed for the night and supped of a piece of fish, a basin of tea, and a slice of bread. The chief factor and the botanist slept in the canoe pulled up on the shore, while the natives, who had come all the way with nothing to eat except salmonberry shoots, spent the night in roasting and eating a twenty-six pound sturgeon. All of it had disappeared by daybreak, when the journey was resumed. Fort Vancouver was reached late that night.

Douglas was enraptured with the grand scenery along the river and round about. It was much more magnificent than that of his native land. He was drawn particularly by the great snow-clad peaks: Hood, Jefferson, and St. Helens. "The scenery round this place is sublimely grand," he wrote, "lofty, well-wooded hills, mountains covered with perpetual snow, extensive natural meadows, and plains of deep, fertile, alluvial deposit, covered with a rich sward of grass, and a profusion of flowering plants." [2]

Making Fort Vancouver his headquarters and living first in a tent, then in a deerskin lodge, and later in a hut of cedar bark, he spent the first six months in plant collecting along the Columbia River between its mouth and the Great Falls (Celilo) 220 miles up, and on the Multnomah (Willamette) River. His aim was to get as large a collection as possible to send home on

[1] Douglas, *Journal*, p. 106.
[2] Douglas, "Sketch," *OHQ*, V, 243.

the *William and Ann,* which was to leave in October after a trip northward along the coast. (He had some thought of taking this trip but was advised against it by McLoughlin owing to the disturbed condition of the natives.)

Back and forth along the river he went, sometimes with an Indian guide, at other times with the Company's hunting or trading parties or its fur brigades, but often alone. Scouler accompanied him on a few journeys, having come up from Fort George while the ship was unloading. He soon got accustomed to the hardships of wilderness travel. As he could not always take a tent, he sometimes slept on the riverbank under an upturned canoe. More frequently he had no shelter except the branches of a pine tree, an experience which at first seemed peculiar: "Each individual takes his blanket and with all the complacency of mind that can be imagined throws himself on the sand or under a bush just as if he were going to bed." [3] The makeshifts on the Highland tours with Hooker had been trivial compared with these. However, being obliged to put up with them, his tune changed: "The luxury of a night's sleep on a bed of pine branches can only be appreciated by those who have experienced a route over a barren plain, scorched by the sun, or fatigued by groping their way through a thick forest, crossing gullies, dead wood, lakes, stones &c." [4]

For food he was obliged to depend largely on birds that he shot and on fish that he caught himself or bartered from the natives. He became an expert shot, as the following occurrence shows. On his way back from a journey down the river he was entertained by Cockqua, a friendly chief on the north side whose tribe was at war with the Clatsops on the south. Entertainment began with supper from an immense sturgeon ten feet long weighing four or five hundred pounds—"the most comfortable meal I had had for a considerable time." In the evening about three hundred warriors danced the war dance and sang

[3] Douglas, *Journal,* p. 107.
[4] *Ibid.,* p. 128.

death songs, "which to me alone imparted an indescribable sensation." An invitation to sleep in the Chief's lodge Douglas declined because of the numerous fleas and because he did not want to show fear of the enemy, so he spent the night in his own tent. In the morning one of the best marksmen of the tribe exhibited his skill with the bow and arrow and also with a gun. Finding that the Indians had no idea of shooting on the wing, Douglas raised an eagle by throwing a stone at it and then calmly shot it down. This caused great surprise among Cockqua's warriors. The Indian marksman then offered Douglas a shot at his hat thrown into the air. The botanist shot away the whole crown, leaving only the brim. Thereupon his fame was sounded throughout the camp and high offers were made for his gun, while Chief Cockqua declared: "Cladsap cannot shoot like you." After this experience Douglas found it to be "of the utmost value to bring down a bird flying when going near the lodges, at the same time taking care to make it appear as a little thing and as if you were not observed." [5]

He carried on his plant collecting under great difficulties. During the early part of the summer the almost continual rains caused the loss of many of his specimens, for in spite of oilcloths he was unable to keep his plants dry or to preserve any birds. Before he could lie down to sleep it took an hour to dry his blanket. Sometimes he ran short of food and was so weakened by hunger and fatigue that he could only drag himself about. On one occasion he was so tired on making camp that he fell asleep while cooking two partridges he had caught, and on awakening at daybreak he found his supper burned to ashes and three holes in the bottom of his kettle. He managed to make some tea, however, which he declared "the monarch of all food after fatiguing journeys," by scouring out the lid of his tinder box to boil the water in.

The change from the wet climate of the lower Columbia to the arid regions of the interior was equally distressing. A nine-

[5] Douglas, *Journal*, p. 138.

teen-mile walk across a barren plain at 97°F. in the shade without a drop of water caused much suffering; at night his feet burned like the hot sand, and the upper portions were in big blisters. Nevertheless, it was here that he discovered the beautiful *Clarkia pulchella* and the gaudy mariposa lily (*Calochortus macrocarpus*), both of which he introduced into England.[6]

The natives, too, were troublesome. Like children, they were inquisitive and took up much time with questions and palaver. Often they were treacherous and would steal or murder if they thought they would not be punished. Until a few years previous they had been quite hostile, frequently pillaging the Company's boats at the Cascades and Dalles[7] portages and killing some of the men. However, as they became accustomed to the firm and just policies of the Company in its trade relations, their hatred and resistance lessened. Douglas's journey to Celilo in September, 1825, was the first made without an armed guard. Contact with the whites led the tribes to become friendly and imitative; one of his guides was such a truckler that he had been given a name meaning "yes."

They placed strange interpretations on the white man's inventions. When McLoughlin in June each year set up on the river's edge a gauge to measure the high water of the summer freshets, they concluded that it was to stop the river from rising further.[8] When Douglas made an effervescent drink his canoemen and guides were surprised to see him drink it "boiling," as they thought. For this, and for lighting his pipe with the sunlight and a magnifying glass, they classed him with the bad spir-

[6] Lewis and Clark found the *Clarkia pulchella* but made known only dried specimens (named after Clark by Pursh); Douglas sent home seeds. Other Clarkias introduced by Douglas are *C. elegans* and *C. rhomboidea*, the latter from California. Three other mariposa lilies were also introduced by Douglas from California.

[7] *Dalle* is French for "slab" or "flagstone," descriptive of the flat basaltic rocks over which the river flows rapidly through a narrow channel.

[8] W. F. Tolmie, in *Transactions of the Oregon Pioneer Association for 1884* (Salem, 1885), p. 37.

its and called him *Olla-piska,* the Chinook words for fire. But beyond all their comprehension was his putting spectacles on his nose; at this they clapped their hands over their mouths in astonishment and fear. Afterwards, as they came to know him better and to understand his motives in gathering plants, shrubs, and grasses to send to England, they regarded him as one of King George's chiefs and gave him the title of the "Grass Man."

He continued his work undaunted; he had a robust constitution and a merry heart;[9] fresh difficulties brought forth fresh enthusiasm and determination, and few seemed unsurmountable. He even considered climbing Mount Hood, but when he approached within a few miles of it the barriers appeared too great. He was to try again later.

Difficult and exhausting as were his journeys, he was happy in the rich harvest of plants. His expectations were more than realized. During the six months following his arrival he collected 499 species, and as far as possible obtained from twelve to twenty-four samples of each.[10] These he carefully dried in paper brought from England for the purpose, and he classified and minutely described each species in his diary. This work he did while resting between journeys. Page after page of his diary was taken up with the detailed descriptions; the same is true of his journal as printed.

Not all his species were new to botanists, but a large number were. Among these were several varieties of lupine, penstemon, evening primrose (*Oenothera*), and mimulus, the beautiful blazing-star (*Mentzelia albicaulis*), the wild hyacinth (*Brodiaea grandiflora*), and the pretty little Blue-eyed Mary (*Collinsia grandiflora*); while among the shrubs were the dainty ocean-spray (*Spiraea*), the attractive arbutus (or madrona), the shiny evergreen Oregon grape, the ubiquitous salal, and

[9] George Barnston, "Abridged Sketch of the life of Mr. David Douglas, Botanist, with a few details of his travels and discoveries," *Canadian Naturalist and Geologist,* Montreal, V (1860), 124.

[10] Letter, Oct. 3, 1825, Douglas to W. J. Hooker, Kew Gardens.

various honeysuckles, berries, and currants. Among the woods were the broad-leaved maple and the vine maple, the latter of which Douglas said was "called by the voyageurs *Bois de diable* [wood of the devil] from the obstruction it gives them in passing through the woods." In some cases the obtaining of seeds was a difficult matter. Nevertheless, all of these plants, and many others, he had the distinction of introducing to Europe.[11] A lot of them he named, but some of the names have not been retained because they were not published, or have dropped into synonymy with earlier published names for the same entities.[12]

Amusing incidents occurred in connection with some of the plants. There were the salal berries which he put in alcohol for sending to England, but which were stolen at Fort Vancouver "by some evil disposed person . . . for the sake of the spirits they were in." There was the camas (*Camassia esculenta*), an onion-like root which was baked by the Indians and formed a substantial item of their food, but which caused flatulence to such an extent that once when in an Indian hut Douglas "was almost blown out by strength of wind." There was the tobacco plant (*Nicotiana*) which he took from an Indian garden without the owner's permission. To appease his wrath he presented him with two finger-lengths of the European weed, whereupon the Indian genially described the method of cultivating the native variety. "His knowledge of plants and their uses gained him another finger-length. When we smoked we were all in all."

Among his discoveries was the musk-scented monkey-flower (*Mimulus Moschatus*),[13] which was soon naturalized in Great

[11] The Royal Horticultural Society's list of "Plants Introduced by David Douglas during the Years 1826-34" contains over two hundred plants, about a third of which are from California. See Douglas, *Journal*, pp. 326-336. A condensation of this list is contained in Appendix A.

[12] At least sixty-two plants of British Columbia (including many he did not introduce) still have the names he gave them (list given the author by J. W. Eastham, plant pathologist, British Columbia Department of Agriculture, Vancouver, B. C.).

[13] Edwards's *Botanical Register*, London, XIII (1827), fig. 1118.

Britain and became very popular.[14] It has been of particular interest to botanists because it has lost its scent—a phenomenon which so far has not been fathomed.[15]

His attention was not confined to vegetation. At the Cascades specimens of birds and animals were obtained and the petrified trees wondered at. There, too, was the great salmon fishery of the Indians, who gathered from far and near during the fishing season. The price amazed him. He bought two salmon, weighing about thirty-five pounds each, for two inches of tobacco (½ ounce; value, two pence). In England they would have cost him three or four pounds sterling.

The dense, dark forests of evergreen trees might be depressing to some, but not to Douglas. To wander through the shady stillness among the tall, erect trunks was a privilege and a pleasure. The fir which he knew as *Pinus taxifolia,* but which was later named after him, he found nearly everywhere.[16] The immense size of these trees amazed him. He measured some fallen ones and found that the largest was 227 feet long and 48 feet in circumference three feet above the ground.[17] Their great height proved an obstacle to obtaining the cones. His buckshot would not reach them; the trees were far too large to cut down with his hatchet; and, as for climbing, he had "already learned the propriety of leaving no property at the bottom of a tree." [18] It was some time before he succeeded in getting cones.

The gigantic size and compact uniformity of the tree led him to call it "one of the most striking and truly graceful objects in Nature." He anticipated its future commercial value, saying: "The wood may be found very useful for a variety of domestic

[14] F. J. Chittenden, "The Lost Scent of Musk," *Gardener's Chronicle,* LXXXVIII (1930), 457.

[15] *Gardener's Chronicle,* XLV (1909), 267; LXXV (1924), 79; LXXXVIII (1930), 259, 349, 399, 457, 520-521.

[16] *Pinus* was the generic name for all pitchy trees in his day.

[17] Douglas, *Journal,* p. 339.

[18] A reference to the Negro incident at Amherstburg (see Chapter 3).

purposes: the young slender ones exceedingly well adapted for making ladders and scaffold poles, not being liable to cast; the larger timber for more important purposes, while at the same time the rosin may be found deserving attention."[19]

On his journey up the Willamette—he went twenty-four miles above the falls (now Oregon City) to a Calapooie village —he found that the Indians carried in their tobacco pouches some pine seeds which they ate as nuts. He learned that they came from an unusually large tree growing far to the south. As the Indians had no perfect seeds, Douglas asked one of the fur hunters, Jean Baptiste Desportes McKay, to obtain some for him. This was the inception of Douglas's search for the sugar pine, which will be recounted in a later chapter.

Another journey fruitful in pines was one to the mountains near the Cascades. A most difficult ascent was made on the north side of the river; so difficult that his guide, Chumtalia, an Indian chief, avoided it by pleading illness, and in his place sent his young brother and two other natives. After a shivering night on the summit without blankets—only a fire—and with scarcely any food, he ascended the south side, which was less difficult, yet required fifteen hours' steady climbing. Here he was rewarded by finding two new firs, which he thought "the grandest trees of the tribe" and named *Pinus nobilis* and *Pinus amabilis,* now known as *Abies nobilis* and *Abies amabilis.* Also found here were the bear grass (*Xerophyllum asphodeloides*), called *Quip Quip* by the Indians and used by them for making watertight baskets, and new species of rhododendron and pine drops (*Pterospora*).[20]

With regard to the naming of the firs, Carl Hansen, an authority on the subject has written as follows:

It is unfortunate, and it seems unjust that the discoverer of an object in natural history—one, who, like Douglas, has the energy and

[19] Douglas, *Journal,* pp. 339-340.

[20] Douglas appears to have been the first to ascend any of the Cascade Mountains and the first to call them by that name.

daring to explore, the intelligence to comprehend when he has an object in sight that is new to science, and, moreover, the ability to describe and name it correctly, referring it to the proper genus in vogue at the time of publishing—it seems unjust that such a namer should subsequently lose the honours of discovery and authorship, because, forsooth, another view of the relative importance of groups places the object in another category, and therefore another person, to wit, the one who so places it, becomes the author of the species. Such is the latest usage, however, based upon lately revised ancient laws of nomenclature; and, in the long run, it works less mischief than would a reverse rule, whereby pseudo-scientists could air their vanity by foisting upon us a host of unfounded terms at will.[21]

For a long time the existence of the *Abies amabilis* was doubted by botanists. Although discovered by Douglas in 1825, and reported by John Jeffrey, collector for the Scottish Oregon Society, from the Fraser River region in 1851, other botanists who visited northwest America did not see it. For several years authors catalogued it, relying on the reports of Douglas and Jeffrey, but at length faith in these reports died out, and botanists came to the conclusion that the traditional *amabilis* was either a form of some other species or a mythical fir formed by mixing specimens of two or more species.

Vindication of Douglas's claim did not come for many years. Hansen thus relates it:

In 1879, Dr. Engelmann, who was elaborating the *Abietineae* for the 'Botany of California,' boldly declared that there must have been some mixing of Douglas's and Jeffrey's specimens, and the 'Lovely Fir' was therefore a myth—a fictitious species.

The next year, however (in 1880), the Doctor, accompanied by Professor Sargent and Dr. Parry, made an extended exploration of the forests of the great Northwest, intent upon settling, once for all, several doubtful subjects that had all along marred our descriptive work, and they were very successful—after toiling as Douglas had done, finding that in every case the original explorers had told but the truth.

[21] Carl Hansen, "Pinetum Danicum," *Journal of the Royal Horticultural Society*, London, XIV (1892), 457.

On Silver Mountain,[22] near Fort Hope, Fraser River, at an elevation of 4,000 to 5,000 feet, they came in sight of a beautiful, unfamiliar Fir, which they at once recognized as the long-lost "Amabilis"—the same that Douglas had first made known fifty-five years previously.

A few weeks afterwards Professor Sargent ascended the very mountain where Douglas made his fifteen-hour climb just south of the cascades of the Columbia River, and there also was found the Amabilis Fir in all its pristine beauty, and not far distant the Noble Fir, as Douglas had portrayed it.[23]

Another tree that Douglas made known is the Sitka spruce. It thrives especially well in the humidity of the North Pacific Coast, and is valuable in pulp and paper making and airplane construction. The first seeds that he got were lost, so it was not until a later visit that he was able to send any home. He called it *Pinus Menziesii* in honor of his predecessor, but the present name is *Picea sitchensis.* He also had to wait until later to obtain seeds of the western white pine, which he found on a journey in the vicinity of Mount St. Helens. He introduced it as *Pinus monticola,* its present name.

Two whole weeks were taken in packing and labeling the plants and specimens he had collected for shipment home on the *William and Ann.* Besides sixteen large bundles from the Columbia there were eight bundles of plants obtained at the various stopping places during the journey from England. There was also a box of birds and quadrupeds, and another containing Indian handiwork. Most important of all was the collection of seeds—over a hundred varieties—which were packed in a chest by themselves. Where they could be divided, a small portion was kept by Douglas to be taken or sent home across

[22] Half a mile east of Isolillock (Holy Cross) Mountain. Not the Silver Peak that is several miles south.

[23] Hansen, "Pinetum Danicum," p. 457. See also George Engelmann's letter in *Gardener's Chronicle,* XIV (1880), 720, his "Synopsis of American Firs" in *Transactions of the Academy of Sciences of St. Louis,* III (1868-1877) 593-602, and his "Notes on Western Conifers" in the *Botanical Gazette,* Indianapolis, VII (1882), 4, 5.

the continent, as he had been instructed—evidently a precaution against the loss of those sent by ship.

Dr. McLoughlin's kindness was not overlooked. Douglas had the distinction of being the first white visitor to the Hudson's Bay Company's posts on the Columbia, and McLoughlin played the part of host most nobly. Always attentive to the young man's needs, he advised him as to itineraries and guides, and placed at his disposal horses, canoes, and men whenever they could be spared. This was characteristic of the Father of Oregon, whose good will toward his guests has become well known.[24] He also sent a general notice on behalf of Douglas to the posts in the interior, so that the botanist got assistance wherever he went, with all the comfort that could be afforded. In recognition of his kind attitude, McLoughlin, like Captain Hanwell, was presented with the Horticultural Society's large silver medal.[25]

[24] "The very Christ of Northwest occupation" (Hubert Howe Bancroft, *History of British Columbia*, San Francisco, 1890, p. 305).

[25] Presented between May 1, 1826, and May 1, 1827 (*Transactions*, vol. VII, 1830, p. 2.

6

WITH CHIEF THA-A-MUXI

A FALL on a rusty nail while packing the last of the boxes for shipment to England caused a large abscess on Douglas's knee joint and a badly inflamed leg. This laid him up for nearly three weeks and prevented his going to Fort George to see the boxes safely placed in the brig, as he had intended. He wrote Captain Hanwell, therefore, requesting that they be placed in an airy situation, particularly the chest of seeds, and above water level if possible.

Contrary winds detained the vessel at the mouth of the river for about two weeks. Learning of this, and the condition of his knee having improved, he decided to visit his old shipmates and then proceed north to Whitbey Harbour (now Gray's Harbor) in search of new plants and seeds. Leaving Fort Vancouver on October 22 with Alexander McKenzie and four Indians, he hurried down the river, but was delayed by having to repair the canoe, after striking a snag which split the craft from end to end and swamped them. Heavy swells lower down the river added to the delay, and on arriving at the mouth he was disappointed to learn that the ship had sailed just an hour before.

Staying overnight with Concomly, the great chief of the Chinooks, Douglas met the chief's brother, Tha-a-muxi, or "The Beard," who lived at Gray's Harbor, and arranged to accompany him there.

Tha-a-muxi was an unusual character—"a fine old man," Douglas calls him. While they were being ferried across the mouth of the Columbia in Concomly's big war canoe manned by twelve of his braves, a sudden storm arose which drove the sea over them repeatedly and washed away nearly all their food. They landed wet and miserable. Thinking that a little stimulant would be much appreciated, the Grass Man offered the other chiefs "a dram of well-watered rum, which pernicious liquor they will, generally speaking, make any sacrifice to obtain." Tha-a-muxi, however, proved to be an exception to the rule. He declined the liquor.

> On my enquiring the reason of his temperance, he informed me that some years ago he used to get drunk, and become very quarrelsome; so much so, that the young men of the village had to take and bind him hand and foot, which he looks upon as a great disgrace, and will taste spirits no more. In lieu of drinking, however, I found him an expensive companion, from his addiction to tobacco. So greedily would he seize the pipe and inhale every particle of smoke, that regularly five or six times a day he would fall down in a state of stupefaction. In self-defence I was obliged to smoke, when I found that my mode of using the Indian weed diverted my companion as much as his had me. "Oh," cried he, "why do you throw away the food? (smoke). See, I take it in my belly." [1]

The journey turned out to be a terrible one. After portaging four miles across Cape Disappointment and dragging the canoe through the woods and over rocks, stumps, and gullies to Willapa Bay, they met with thick fog and rain. Douglas's knee became stiff and painful, owing to the exertion and the cold and wet. At Cape Shoalwater (erroneously called Foulweather by Douglas) they were stormbound for two days by high winds and rain, sleet, and hail, with scarcely any fire and no food except a little chocolate, a few kinnikinnick berries, and some roots of *Sagittaria* (arrowhead, a marsh plant) and *Lupinus littoralis* (*somuchtan*, in Chinook), a wild licorice which Douglas found nutritious and wholesome. Confronted with a portage of sixteen miles to Gray's Harbor, Douglas sent the canoe back to the Columbia, two of the Indians being glad to take it when they found all the provisions were gone. The weather became so violent it was necessary twice to shift the camp back from shore because of the rising sea.

[1] Douglas, "Sketch," *OHQ*, V, 254.

In misery Douglas dragged himself over the portage to the south side of Gray's Harbor, on his legs from four in the morning until six in the evening, with no food for two days and his knee tormenting him. "I hardly can give an idea of my afflicted state."

The day was followed by more misery: a wet, blustery night spent shivering over a poor fire in an improvised shelter made of pine branches, grass, and a few old cedar-bark mats. "The following day found me so broken down with fatigue and starvation, and my knee so much worse, that I could not stir out."

Luckily, toward dusk, when the weather moderated, he managed to crawl out with his gun, and "providentially," as he thought, "killed five ducks with one shot, which, as might be expected, were soon cooked; one of the Indians ate a part raw, the other did not take time to pluck the feathers off but literally burned them to save time. I was certainly very hungry, but as soon as I saw the birds fall my appetite fled; it had brought such a change over me that I could hardly persuade myself I had been in want. I made a basin of tea, on which, with a bit of duck, I made a good supper. Very little sufficed me." [2]

Their fire having drawn the attention of Tha-a-muxi's people across the bay, a canoe was sent for them. Reaching the chief's village on the Chehalis River (in the vicinity of the present city of Aberdeen), he rested and recuperated for several days as the chief's guest—the first white man to receive such kindness from the Chehalis Indians. After the exhausting experiences of the past week comfort of the crudest kind was appreciated, so he was in no mood to find fault with the dirt, filth, and stench in which his hosts lived—so bad that John Work, a fur trader accustomed to Indian life, who passed through a year earlier, was surprised that human beings could exist amid such conditions.[3] Instead of complaining, Douglas says he "had every kindness and all the hospitality Indian courtesy could suggest."

He botanized in the vicinity and obtained some new seeds, in-

[2] Douglas, *Journal*, p. 148.

[3] "Journal of John Work, November and December, 1824," T. C. Elliott, ed., *Washington Historical Quarterly*, Seattle, Wash., III (1912), 206-207.

cluding the interesting bear grass and wild licorice. He then started up the Chehalis (called Cheecheeler by Douglas, "a large stream, nearly as large as the Thames, very rapid, with numerous cascades") intending to reach its source, but after going sixty miles the rainy season and the difficulty of getting the canoe up cascades and through shallow water led him to give up the idea. Instead of returning as he had come, he decided to cross over to the Cowlitz River and return by it to the Columbia.

Tha-a-muxi, or The Beard, had accompanied him up the Chehalis. On leaving to return home he entreated Douglas to shave him, so he "might look more like one of King George's chiefs." [4] As Douglas thought the chief "had pretensions to civilization and aped with nicety European manners," he not only complied but also "invited him to come at the New Year to see me, when I would give him a dram, a smoke, and shave him again. He told me before he left, to let all King George's chiefs know of him, when I spoke to them with paper." [5]

Leaving the Chehalis, Douglas found the trail in bad condition on account of the heavy rains, and three of the streams crossed were so high that he had to take off his clothes and swim. His small quantity of baggage was carried on a horse hired from "the most mercenary rascal" he had ever seen, who charged him "twenty shots of ammunition, two feet of tobacco, a few flints, and a little vermilion." And the biggest part of the horse's burden was its mercenary owner.

Arriving at the Cowlitz, he obtained a boat from Chief Schachanaway and proceeded down the river and up the Columbia, using his blanket and coat as sails. He reached Fort Vancouver on November 15 after an absence of twenty-five days, "during which I experienced more fatigue and misery, and gleaned less than in any trip I have had in the country."

On November 18 the annual express of the Hudson's Bay

[4] Letter, April 3, 1826, Douglas to Scouler, printed in Douglas, "Sketch," *OHQ*, V, 333.

[5] Douglas, *Journal*, p. 149.

Company, consisting of two boats and forty men, came from Hudson Bay. The yearly arrival of people and news from across the continent was a great event for the fur-trading posts on the Columbia; and this year's express was the first to terminate its journey at the new establishment. An eager welcome was given, in which Douglas joined, since he expected letters from England. He was "heavily disappointed," however, upon receiving nothing, for the express had left the Bay before the arrival of the ship from England; and "it was tantalizing to reflect that whatever might have been sent to me by that vessel, must now lie on the other side of the great Continent of America until November of next year." [6]

Alexander Roderick McLeod, who was in charge of the express, had spent five years in the Mackenzie River country and he gave Douglas much information about it. According to the natives, the long-doubted Northwest Passage did exist.[7] Douglas saw in McLeod "a great example of perseverance, visiting the Polar Sea, the Atlantic and Pacific Oceans, in the short space of eleven months."

The activities of a plant hunter must of necessity be confined chiefly to the period when the plants are in flower and the time when the seeds are ripe. It was then that Douglas worked with such tremendous zeal. But during the off-seasons he was by no means idle. During this first winter on the Columbia he collected samples of wood, gathered mosses, made up collections of preserved animals and birds with detailed descriptions, and kept a journal of the weather.

The day before Christmas a torrential rain inundated his cedar-bark hut to a depth of fourteen inches, so on Christmas Day, at Dr. McLoughlin's kind invitation, he took up quarters in the latter's half-finished house. After the customary service of the morning (conducted by the Doctor), the gentlemen of

[6] Douglas, "Sketch," *OHQ*, V, 258.

[7] Letter, March 24, 1826, Douglas to W. J. Hooker, printed in Douglas, "Sketch," *OHQ*, V, 330.

the Fort went horseback riding, but his knee prevented Douglas from joining them.

New Year's Day, 1826, brought this dismal entry in his journal:

> Commencing a year in such a far removed corner of the earth, where I am nearly destitute of civilized society, there is some scope for reflection. In 1824, I was on the Atlantic on my way to England; 1825, between the island of Juan Fernandez and the Galapagos in the Pacific; I am now here, and God only knows where I may be next. In all probability, if a change does not take place, I will shortly be consigned to the tomb. I can die satisfied with myself. I never have given cause for remonstrance or pain to an individual on earth. I am in my twenty-seventh year.[8]

[8] Douglas, *Journal,* p. 152.

7

UP THE COLUMBIA TO KETTLE FALLS

DOUGLAS had been instructed to return to England in 1826, but whether by ship around Cape Horn, or by land to Hudson Bay and then by ship, had been left for him to decide, although Sabine favored the latter.[1] He had not been long in the country before he decided upon the overland route. "The last effort of existence shall be used to accomplish the latter," he wrote Hooker, and knowing the professor's great interest in mosses and liverworts (scale mosses) he added: "Is there any fine *Musci* or *Jungermanniae* on those high spots?" [2]

To return that way would mean leaving Fort Vancouver in March by the annual express for Hudson Bay. That is what one would expect to find him doing, in view of his gloomy reflections of New Year's Day. Those meditations passed, however, and as the days lengthened brighter thoughts possessed him—and a fresh determination. Successful as he had been, he realized nevertheless that the field had by no means been covered. The vast territory far up the Columbia and its tributaries toward the Rocky Mountains remained to be explored.

Convinced that his work was unfinished, he decided to stay for another year. Although it was contrary to his instructions, he felt that in justice to the Society's best interests he could not do otherwise. Asking pardon for his disobedience, he pointed

[1] Letter, July 26, 1825, Joseph Sabine to Douglas, copy at Royal Horticultural Society.
[2] Letter, Oct. 3, 1825, Douglas to W. J. Hooker, Kew Gardens.

out, first, that he was incurring very little expense and, second, that his illness during the preceding autumn probably had meant the loss of many interesting seeds. He added: "Lest the former should be made any objection to, most cheerfully will I labour for this year without any remuneration, if I get only wherewith to purchase a little clothing." Knowing that he could not receive an answer until long after the year had passed, possibly not until his arrival in England, he hoped for the best—and doubled his efforts.

In deciding to remain he faced a difficulty even greater than the lack of clothing. His eyesight, which always had been weak, was becoming much impaired. Without pain or inflammation, a dimness came over his eyes which was a great handicap, especially when shooting, since he often depended on his gun for food and for obtaining seeds and branches in his plant collecting.[3] But his enthusiasm won the decision; so, after packing up his recent collections of plants and birds to be sent to England by the next ship and getting ready the duplicate set of seeds (197 papers in a tin box) to go across the continent, he left Fort Vancouver with the Hudson's Bay express on March 29, 1826—not en route home, but on what was to be his first extended journey up the Columbia. With the kind permission of Dr. McLoughlin he was enabled to take with him, for packing the plants he anticipated obtaining, thirty quires of paper weighing 102 pounds, "an enormous indulgence" in view of the difficulties of transportation, but "rather than go unprovided in this respect I curtailed the small supply of clothing," taking only two shirts, two handkerchiefs, blanket and cloak, one pair of shoes but no stockings. "Thus I adapt my costume to that of the country."

The two boats comprising the express must have been well loaded, for besides seventeen men and the freight they carried three calves and as many pigs for the new post being estab-

[3] Letter, April 3, 1826, Douglas to Scouler, *Edinburgh Journal of Science,* VI (January, 1827), 117.

lished at Kettle Falls[4] (Fort Colville).[5] It was the first livestock to be shipped into the interior, being the offspring of pigs brought from Hawaii and of cattle from California.[6]

On reaching the Dalles portage the party had to be on guard all night against four or five hundred Indians who were indignant at the quantity of tobacco received in exchange for salmon and were threatening trouble. At the Great Falls (Celilo), always a danger point, the natives were more offensive and tried to pillage the boats. No doubt the calves and pigs were a great temptation. As John McLeod, who was in charge of the express, pushed one of the Indians back, another aimed an arrow at McLeod. Douglas seized his gun loaded with buckshot and threatened to shoot if the arrow was fired.[7] At this tense moment a tall Cayuse chief friendly to the whites stepped forward and settled the matter in a few words.

To reward the chief, Douglas, as one of King George's chiefs (the Grass Man), bored a hole in the only shilling he had, which had been in his pocket ever since leaving London, and suspended the coin by a bit of wire to the Indian's nose, the septum of which was already perforated. This ceremony established a seal of friendship between them and the chief promised that the whites should not be molested. Nevertheless, sleep was again out of the question, so Douglas occupied himself in writing a letter to Dr. Hooker telling of his discoveries and his future

[4] "The Journal of John Work," T. C. Elliott, ed., *Washington Historical Quarterly*, V (1914), 284.

[5] Although the post was named after Andrew Colvile, a director of the Hudson's Bay Company, the spelling Colville also came into early use and has since been adopted.

[6] "Scouler's Journal," p. 166.

[7] Malcolm McLeod, ed., in *Peace River. A Canoe Voyage from Hudson's Bay to the Pacific by the late Sir George Simpson (Governor Hon. Hudson's Bay Company) in 1828. Journal of the late Chief Factor, Archibald McDonald* (Ottawa, 1872), p. 108, and Hubert Howe Bancroft in his *History of the Northwest Coast* (San Francisco, 1884), II, 469, both erroneously credit the saving of McLeod's life to James Douglas. James Douglas did not arrive on the Columbia until June (see below).

plans. The sugar pine (*Pinus Lambertiana*), seeds of which had come to his notice while up the Willamette, and a cone of which had since been brought to him by Jean Baptiste Desportes McKay, he enthusiastically described to Hooker as "the most princely of the genus, perhaps even the grandest specimen of vegetation."

From the Falls to the Spokane River the journey was in the dry, barren, sagebrush country. Throughout this distance (532 miles) he botanized diligently. While the boats were being paddled or lined [8] slowly upstream against the strong current, he walked along the banks and climbed the bordering hills looking for plants, although his knee still troubled him. His efforts were rewarded by finding several new specimens, including the beautiful mission-bell (*Fritillaria pudica*) and some wild parsleys, mustards, and currants. His curiosity was aroused by the humble sagebrush (*Artemisia*) and its several varieties, at least sixteen in number.

At Fort Walla Walla[9] he espied the Blue Mountains in the distance. Thinking that they would yield him most of the Rocky Mountain plants, he arranged with Chief Trader Samuel Black for a journey there in June.

Day after day up the river the party toiled, Douglas clambering along the shore searching for plants, shrubs, herbs, animals, birds, even rocks—anything of interest in the realm of natural history. Through the dangerous Priest's Rapids they went, then on past the site of the present city of Wenatchee at the mouth of the river of that name (called Piskahoas in Douglas's day). At the great southerly bend of the river, where the Okanagan

[8] That is, pulled with ropes by men walking along the bank.

[9] Fort Walla Walla, often called Fort Nez Perces after the tribe of Indians, was situated on the east bank of the Columbia a half-mile above the Walla Walla River, near where Wallula now is. It was established in 1818 by the North West Company. The name means "small, rapid river" (Edmond S. Meany, *Washington Historical Quarterly*, vol. IX, 1918, pp. 204-205; vol. XIV, 1923, p. 132). See also Samuel Parker, *Journal of an Exploring Tour beyond the Rocky Mountains . . . 1844* (Ithaca, N. Y., 1844), pp. 130-132.

River joins it, they tarried a few days at Fort Okanagan[10] (near the present town of Brewster), but deep snow prevented any botanizing. Proceeding easterly they passed the head of the Grand Coulee, where now the greatest dam in the world is situated.

At the junction with the Spokane River, where they arrived on April 11 and rested for a few days, Douglas bade good-by to the express. The calves and pigs continued on to the new establishment at Kettle Falls to give origin to the beef, pork, and dairy business of the Inland Empire, as eastern Washington is called;[11] Douglas's seeds continued on to Hudson Bay and England, to bloom and make famous in Europe the botanical treasures of northwest America. It was a notable cargo that the express carried from Fort Vancouver that year.

Douglas sent only seeds across the continent. He was unable to send dried plants because they were too bulky and would have had to be shipped in chests lined with tin or other metal to keep them dry, which was out of the question since such metals were not obtainable.

That year's express had a particularly difficult crossing of the Rocky Mountains, through "snow from 16-30 feet deep . . . Oft no fire: half of the time in deluge of rain," [12] and McLeod was obliged to cut up his leathern trousers to make snowshoes.[13]

At the Spokane River, Douglas met John Warren Dease, the

[10] The name "Okanagan" has the distinction of having been spelled in no less than forty-five different ways. See Leonard Norris, "Some Place Names," *Sixth Report of the Okanagan Historical Society, 1935* (Vancouver, B. C., 1936), pp. 133-136, and Symons, p. 130. The name is now spelled Okanogan in the United States, Okanagan in Canada.

[11] T. C. Elliott, "Journal of John Work, Dec. 15th, 1825, to June 12th, 1826," *Washington Historical Quarterly,* V (1914), 284; *Ranald MacDonald, the Narrative of his Early Life on the Columbia under the Hudson's Bay Company's Regime,* Wm. S. Lewis and Naojiro Murakami, editors (Spokane, Wash., 1923), pp. 103-104.

[12] John McLeod, "Journal of Voyage by boat to Boat Encampment—Western end of Athabaska Pass—thence through the Pass, etc." (1826), MS., Dominion Archives, Ottawa. This journal is written in pencil and much of it is illegible.

[13] H. H. Bancroft, *Northwest Coast,* II, 468.

Hudson's Bay Company's commandant in the interior, and was invited to accompany him to Kettle Falls, ninety miles higher up the Columbia. They got there two days later, Douglas having botanized along the river banks as before. En route the party had to portage around a big rapid, which Douglas, upon learning that it had no name, called Thompson's Rapids, after David Thompson, the first explorer to descend the river from source to ocean. Later it was called Grand Rapids, and it now is known as Rickey's Rapids.[14]

The grand and picturesque scenery around Fort Colville impressed Douglas as the finest he had seen, well worth the long journey of nearly six hundred miles. The waterfall, with its drop of twenty-five feet and compact momentum in high or low water, had the reputation of being the only falls on the Columbia never run by Hudson's Bay Company boats.[15] He spent several days botanizing in the vicinity. To the thirty-three varieties collected on the way up the river he added twenty-five more, one a dogtooth violet (*Erythronium grandiflorum*) which greatly pleased him.[16] Specimens of various birds were also obtained.

As the firelock of his gun needed repairs, on May 9 he left on horseback for Spokane House,[17] about a hundred miles southeast, to obtain the services of Jacques Raphael Finlay, one of the earliest trappers on the Upper Columbia and the only craftsman within hundreds of miles who could work in good steel.[18] At the Barrière (now Methow) River the horses were

[14] Edmond S. Meany, in *Washington Historical Quarterly*, XIV (1923), 44.

[15] Angus McDonald, "Angus McDonald: A Few Items of the West," F. W. Howay, Wm. S. Lewis, and Jacob A. Meyers, editors, *Washington Historical Quarterly*, VIII (1917), 198.

[16] On Douglas's collections of dogtooth violets, see E. I. Applegate, in *Madrono*, Journal of the California Botanical Society, III (Berkeley, Cal., 1935), 70-71, 102-106.

[17] On the north bank of the Spokane River about half a mile above the junction of the Little Spokane River (about ten miles northwest of the present Spokane city).

[18] Barnston, "Abridged Sketch," p. 129.

unsaddled and swum across, the men following. Douglas made two trips swimming on his back, one with his paper and pen, the other with his blanket and clothes, holding them above the water in his hands; his guides made three trips each with the saddles and provisions.

Spokane House was reached on the third day, and a kind welcome was received from old "Jocko" Finlay. He regretted that he had no food to offer, since for six weeks he and his family had been living on roots of the camas and the bitter root (or sandhill rose; *Lewisia rediviva,* which has a starchy and slightly bitter root called *spaetlum* by the Indians) and a cake made from the lichen which grows as a parasite on the pine trees. Douglas's provisions were scanty, consisting of dried buffalo meat, tea, and sugar, supplemented by game he had shot on the way, but he promptly divided them and gave half to Finlay, who enjoyed the best meal he had had for some time. The old man quickly put the gun in order, receiving in payment one pound of tobacco, the only currency that Douglas had. Finlay's poor English—for despite his surname he was very much a French-Canadian—and the Scotsman's poor French made it difficult to obtain all the information that Douglas had hoped for regarding plants and animals, but they managed to discuss wild currant bushes and mountain sheep, Finlay promising to try to obtain specimens of both. Douglas botanized briefly in the vicinity, among his discoveries being the hound's tongue, the wild thimbleberry, and a new tree, the western yellow pine (*Pinus ponderosa*).

An Indian burying ground nearby drew his attention with its curious collections of all the property of the dead placed near their bodies, including war implements, clothing, gambling articles. "Even the favourite horse of the deceased is not spared; it is customary to shoot the animal with a bow and arrow, and suspend his skin, with the hoofs and skull, just above the remains of his master."

He returned to Fort Colville as he had come, again swimming

the Methow River. The cold water and his wet clothes caused a severe pain between his shoulders, which, having no medicine, he tried to work off by walking and perspiration. On arriving at the post the pain and a fever kept him in bed for two days. He doctored himself with Epsom salts and Dover's powder (a compound of ipecac, opium, and potassium sulphate producing perspiration and sleep). On recovering he spent two weeks in plant gathering. One day he walked twenty miles up the Columbia and back; another he went the same distance in the hills to the south—each a forty-mile journey. He also tramped ten miles up the Dease (now Kettle) River while William Kittson, a Hudson's Bay Company man, and two Indians canoed up on an inspection trip, returning with them in the evening. The stream, according to Douglas, "had never before been entered by any European."

Having collected three large bundles of select plants and more than forty varieties of seeds (including *Penstemon Scouleri*, which he named after Dr. Scouler), on June 5 Douglas left "the wild romantic scenery of Kettle Falls" accompanied by Kittson, to return by boat to Fort Walla Walla for his Blue Mountains expedition. It was a wild, swift voyage, for melting snows had swollen the river and accelerated the current. At his Thompson's (Rickey's) Rapids, where the river was greatly agitated by dashing over immense rocks, his Scottish caution showed itself:

Here our boatman, Pierre L'Etang, observed that the water was in fine order for shooting or "jumping," as he called it, the Rapid. Good as this plan appeared to him, I must confess that my timidity would not allow me to remain in the boat. Although I am no coward either *in* the water or *on* the water, and have gazed unmoved, and even with pleasure, on the wildest uproar and tumult of the stormy deep, yet to descend these cataracts by way of sport and where no necessity called for it, I could not resolve to do. Therefore Mr. Kitson and I got out and walked along the rocks. No language can convey an idea of the dexterity exhibited by the Canadian boatmen, who pass safely through rapids, whirlpools, and narrow channels, where by the strength of such an immense body of water forcing its way, the stream, as in the present instance, is lifted in the middle, to a perfect convexity. In such places, where you think the next moment must dash the frail skiff and its burden of human beings to destruction among the steep rocks, these fellows approach and pass over with astonishing coolness and skill, encouraging themselves and one another with a lively and exulting boat song.[19]

At Fort Okanagan he found that his "old friend," John Work, had nicely preserved for him a large female grouse and a male black rock grouse, with eggs of the former. Here, too, he met a party of Hudson's Bay Company officers and men from New Caledonia (now central British Columbia) under Chief Factor William Connolly. Among them were Pierre Chrysologue Pambrun, afterwards an important figure in the fur trade at Fort Walla Walla; Francis Ermatinger, who became treasurer of the provisional government of Oregon; and young James Douglas, destined to play a leading part in the future of the country, first as successor to Dr. McLoughlin on the Columbia, then as superintendent in Vancouver Island and British Columbia, and finally as Her Majesty's Governor of both colonies. The pleasure of meeting such men and receiving their kind attentions was "naturally doubly esteemed in this distant uninhabited country."

Douglas accompanied the New Caledonia brigade to Fort Walla Walla, arriving on June 8, having made the journey down the river from Fort Colville in four days, whereas the journey up had taken more than two weeks. The average rate in descending was 115 miles per day, in spite of delays in putting to shore at several places to gather plants. This gives some idea of the speed at which the Hudson's Bay Company men could travel down the Columbia.

The long days of travel from daybreak until dusk left little time for sleep, so on arriving at Walla Walla, Douglas lay down on the floor of the Indian Hall hoping for a good night's rest; "but was very shortly afterwards roused from my slumber by

[19] Douglas, "Sketch," *OHQ,* V, 344.

an indescribable herd of fleas, and had to sleep out among the bushes; the annoyance of two species of ants, one very large, black, ¾ of an inch long, and a small red one, rendered it worse, so this night I did not sleep and gladly hailed the returning day." He then busied himself in packing up his gleanings made since the shipment of the previous October, amounting to more than a hundred distinct specimens, so that Connolly could take them to Fort Vancouver for shipment to England. A letter to Sabine reporting on his activities accompanied them.

Upon leaving, Connolly generously presented him with twelve feet (over two pounds) of tobacco to assist him in his travels. He was greatly pleased: "This article, being, as it were, the currency of this country, and particularly scarce, will enable me to procure guides and to obtain the cheerful performance of many little acts of service, and it is therefore almost invaluable to me." [20]

That evening an Indian arrived from Fort Vancouver with letters and newspapers for Douglas from England. So engrossed did he become that he had another sleepless night. Next day, having had less than twelve hours' sleep during the past five days, he felt unwell and could do little save mend his shoes. The following few days were spent in plant collecting. His eyes began to trouble him again; the high winds blowing fine sand and the sun reflecting from the barren, sandy soil inflamed and pained them so badly that he could scarcely distinguish anything ten yards away.[21] Nevertheless, his plans for the excursion to the Blue Mountains were not abandoned.

[20] Douglas, "Sketch," *OHQ*, V, 347.

[21] "The heat through the summer is excessive, and high winds prevail and blow the sands about and into everything. By the glare of the sun and the flying sands one's eyes are in a continual state of winking, blinking, and torment, if nothing more serious results" (Lt. T. W. Symons, *Report on an Examination of the Upper Columbia River*, Washington, D. C., 1882, p. 50). Lewis and Clark found the Indians of this area much afflicted with sore eyes and attributed it to the same cause. It proved a blessing to the distinguished explorers, for after their trading supplies had become exhausted, they were enabled to return to civilization only by provisions obtained from the natives in exchange for medicines, especially the much-desired eye-water (*The Original*

Just before starting he had another nocturnal disturbance. "A herd of rats" invaded his rough hewn chamber, ate up all the seeds he had recently collected, gnawed through a bundle of dried plants, and then carried off his razor and soap-brush. Not satisfied with these depredations, one of them came back and was in the act of dragging away his inkstand from near his pillow when he became aroused and with a well aimed shot from his gun, "which is my night companion as well as day," ended the mischief. Evidently the thieves were of the species *Neotoma cinerea occidentalis,* a large, bushy-tailed rat which amuses itself by gathering articles of all sorts around its hole, whence the name "pack rat."

Journals of the Lewis & Clark Expedition, 1804-1806, R. G. Thwaites, ed., New York, 1905, vol. IV, (part 2, pp. 360-365).

8

THE BLUE MOUNTAINS

DOUGLAS was delayed in getting started for the Blue Mountains by negotiations with the guide. "Considerable time was taken up explaining to him the nature of my journey, which was done in the following way: I told Mr. Black[1] in English my intended route, who translated it to his Canadian interpreter, and this person communicated it to the Indian in the Kyuuse language, to which tribe he belongs. As a proof of the fickle disposition and keenness of bargain making in these people, he made without delay strict inquiry what he should get for his trouble. This being soon settled, then came the smaller list of present wants, beginning, as his family had been starving for the last two months, and he going just at the commencement of the salmon season, by asking Mr. Black to allow them something to eat should they call, which was promised. Afterwards a pair of shoes, and, as his leggings were much worn, leather to make new ones was necessary; a scalping knife, a small piece of tobacco, and a strip of red coarse cloth to make an ornamental cap. This occupied two hours and was sealed by volumes of smoke from a large stone pipe." [2]

The language difficulty not only caused delay but also showed the need for taking along an interpreter. For this duty he got the Canadian's son, the Young Wasp.

The route was along the south side of the Walla Walla River, across the South Branch, then south to the foot of the mountains, where they camped the second night. Their food supply con-

[1] Chief Trader Samuel Black.
[2] Douglas, *Journal*, p. 189.

sisted chiefly of pemmican and dried salmon, with a little biscuit, tea, and sugar. Douglas walked to botanize except at stream crossings and bad places. The third day they ascended thirty-five or forty miles in an easterly direction and reached timber line. The following day, starting at daylight as usual, they got to the snow and continued until four o'clock in the afternoon, when "the horses were stopped by deep wreaths of eternal snow, about 1500 feet below the extreme height of the range." Moreover, the guide became troublesome and pretended to be afraid to go farther. This was a great disappointment, for Douglas wanted to cross the summit and descend to the Grande Ronde,[3] the fine large valley on the other side, where he had great expectations of new plants.

Leaving his companions and the horses, Douglas resolutely set out alone on foot with his gun and some paper. At first he found walking on the soft snow without snowshoes very fatiguing, but toward the summit there was a hard crust, and soon he was pleased to stand "on the highest peak of those untrodden regions where never European was before me. The height must be great—7000 to 7500 from the platform of the mountain, and on the least calculation 9000 above the level of the sea."

These estimates are excessive. He probably reached somewhere in the vicinity of Tollgate in northeastern Oregon (Umatilla National Forest). The mountains there do not exceed 6200 feet above sea level.

His admiration of the grand view of the surrounding country was suddenly interrupted by a terrible thunderstorm accompanied by hail and wind, which drove him back down the mountain to join his companions. Arriving wet through, with no fire owing to the high wind and no clothes to change into, he stripped and rolled himself in his blanket to sleep. "Precisely at twelve I was so benumbed with cold that on endeavouring to get up I found my knees refused to do their office. I scoured

[3] So called by the French-Canadian trappers from its fine appearance, being more or less hemmed in by the hills.

them well with a rough towel, and as the storm was over made a cheering fire. I could not resist the temptation of making a little tea, which I found restored me greatly (thermometer 26°). If I have any zeal, for once and the first time it began to cool. Hung my clothes up to dry and lay down and slept until three o'clock."

With the morning his spirits revived and he proposed that they take a less elevated but more circuitous route over the mountain. But his companions would have none of it. As he found out afterwards, the Young Wasp was treacherous. The "rascally" boy misinterpreted Douglas's instructions to the guide, telling him not to go for it was too dangerous. Consequently, the guide urged that to cross the Utalla (Umatilla) River would require swimming or making a raft, and that since his nation was at war with the Snake tribe, in whose confines they were going, in all likelihood the Snakes would steal their horses and perhaps kill the whole party. Seeing the guide's unwillingness and realizing that he could not go alone, Douglas gave up the idea and devoted himself to plant collecting for two days where he was, returning to Walla Walla post on the 24th of June.

Although they had not reached the Grande Ronde, the expedition was not unproductive. Several plants were gleaned, including the *Paeonia Brownii*, the first ever found in America, and a beautiful large golden lupine, which reminded him of the "bonny broom" of his native moors and which he named after Sabine of the Horticultural Society (*Lupinus Sabinii*). There was also the pretty *Lupinaster macrocephalus* (*Trifolium megacephalum*), which he found high up the mountain near the snow.

Somewhat disappointed, he planned a second journey to the mountains, but the guide refused to go, and, when threatened with corporal chastisement, ran away. The Young Wasp had told the guide that Douglas was "a great medicine-man, which is always understood as a necromancer, or being possessed of or conversant with evil spirits, and had the power of doing great wonders; and should he go with me, if he did not do as I wanted, though very likely I did not kill him, he might depend I would turn him into a grisly bear to run and live in the mountains, and he should never see his wife again, which of course acted powerfully on him."[4]

Another guide was obtained, whom Douglas "took the more readily, as he was no smoker, and such a knave that nobody would dare to steal from him."

This time he went along the north bank of the Walla Walla River and up the North Branch, but after traveling for more than two days to the snow line and finding few new plants, he returned. He was tired out, and again suffered great pain from inflammation of the eyes. Only in the mornings could he read or write in comfort. An opportunity arose to send to Fort Vancouver his collections of the past month, so he made a box and packed it with three bundles of dry plants (ninety-seven distinct species), forty-five papers of seeds, three groundhogs (*Arctomys*), and "one curious rat" (perhaps the pack rat he had shot a few weeks before). The box was entrusted to Finan McDonald, well known trader, to take to Fort Vancouver for shipment to England.

Douglas was probably the first white man to ascend the Blue Mountains.[5] Indeed, judging from the attitude of his guides, it is unlikely that the Indians themselves were very familiar with those regions.

It was now July 9. A month earlier, when the Indian messenger had brought Douglas's letters, he had also brought word that McLoughlin had received other letters which he did not wish to risk sending by the Indian, but would send by the interior brigade. Ever since then Douglas had been worrying about those letters. Unable to keep his patience any longer, he now

[4] Douglas, *Journal*, p. 195.
[5] Leslie Haskin, "David Douglas," *Oregon Sunday Journal*, Portland, April 5, 1925, sec. 4, p. 13.

traveled down the river with McDonald's party for two days—subsisting on horseflesh, tormented by mosquitoes (having no tent), and harassed by thieving natives, one of whom he thrashed so soundly with his fists "that he will, I daresay, not forget the *Man of Grass* for some days to come"—in order to meet the brigade under Connolly, who was returning to New Caledonia.[6] They met at the Great Dalles (a few miles below Celilo), and what a happy event it was! After traveling for weeks in the company of Indians it was an indescribable pleasure to meet with friends, "or if even they are strangers, yet the countenance of a Christian is at such times most delightful." His letters "were grasped greedily and eagerly broken open. . . . There is a sensation felt on receiving news after such a long silence, and in such a remote corner of the globe more easily felt than described." So excited was he that he rose four times during the night to read them, and by morning almost knew them by heart.

He returned to Walla Walla on July 15, walking the whole 110 miles along the shore to botanize during the day, and camping with the brigade at night. He found nothing in particular. Six days' travel had been taken in hastening to get his letters.

After arriving at the fort, a party of twenty-eight men under John Work and Archibald McDonald left the brigade in order to go up the Snake River to purchase horses, planning to rejoin the brigade at Fort Colville.[7] Douglas accompanied them. Leaving on July 17, they canoed up the Snake for a week. The heat was so intense, frequently reaching from 98° to 106° F. in the shade, that travel in the open boats was unbearable at midday and had to be confined to morning and evening. Douglas compared the country to the deserts of Arabia, the only difference being the abundance of good water. He obtained relief by bathing frequently. Salmon not being as plentiful or as good as

[6] "The Journal of John Work," *Washington Historical Quarterly*, VI (1915), 27.

[7] *Ibid.*, pp. 31-37.

on the Columbia, the usual fare was horseflesh, boiled or roasted on a stick before the fire.

At the junction of the Snake and Clearwater Rivers (where the City of Lewiston, Idaho, now stands)[8] they found a camp of over five hundred Indians of the Palouse, Nez Perce, and Chamuiemuch nations. A long parley was held to arrange for the barter of goods in exchange for horses "and ended favourably and with great splendour by dancing, singing, haranguing, and smoking." The natives were dressed in their best garments, making altogether a grand and novel spectacle.

As the business was to take some days (the trading slackened as the blankets and beads ran short),[9] Douglas arranged to visit the mountains to the southeast—apparently a continuation of the Blue Mountain range to the southwest which he had visited the month previous.[10] Two days' journey on horseback brought him and his French-Canadian companion, Cock de Lard, to the foothills. Here the guide halted while Douglas pushed on afoot. Nothing new in vegetation was found, but his tramp was not without interest: "Reached the highest peak of the first range at 2 P.M., on the top of which is a very remarkable spring, a circle 11 feet in diameter, the water rising from 9 inches to 3½ feet above the surface, lowering and rising at intervals, in sudden gushes; the stream that flows from it is 15 feet broad, and 2½ deep, of course running with great force as its fall is 1½ foot in 10 and it disappears at the foot of the hill in a small marsh." Douglas could find no bottom at a depth of sixty feet. He named it Munro's Fountain after his friend, Donald Munro, gardener for the Horticultural Society of London.[11] More recently it has been identified as one of the large under-

[8] *Ibid.*, p. 33, note. Called the "Fourches de l'Eau claire" in Barnston, "Abridged Sketch," p. 200.

[9] "The Journal of John Work," *Washington Historical Quarterly*, VI (1915), 34.

[10] The Blue Mountains are now considered to terminate at Snake River Canyon, the present Idaho—Washington boundary, although the mountains to the east are from the same geological uplift.

[11] *Transactions*, Horticultural Society of London, 2nd series, I (1835), 1.

ground outlets of Lake Waha on the slope of Craig Mountain eighteen miles south-southeast of Lewiston.[12] Surrounding the spring he found a large species of wild gooseberry which he named after Munro, but which was afterwards called *Ribes irriguum*, its present name.

Shortage of food compelled a return to camp, where, after traveling all night, they arrived at sunrise. Douglas lay down to sleep, but not for long.

I was hurriedly aroused to take on myself the profession of a soldier, a misunderstanding having taken place between the interpreter and one of the chiefs; the latter accusing the former of not translating faithfully, words became high till at last the poor man of language had a handful of his long jet hair torn out by the roots. On the Indian being reproved, he went off in a fit of rage and summoned his followers, amounting to seventy-three men. All arrived and came to our camp with their guns cocked and every bow strung. As every one of our party had done all in his power that it should be mutually and amicably adjusted and been refused, every one seemed more careless for the result than another. We (thirty-one of us) stood to our arms and demanded if war was wanted; it was answered 'No, we want only the interpreter killed, and as he was no chief there could be but little ill done.' They were told that whatever person we had in our party, whether chief or not, or if it was only an Indian under our protection, should they attempt to kill or disturb him in the least, certainly they would know we had been already in war. The coolness that seemed to be the prominent feature in our countenance had the desired effect of cooling their desire for war and made them glad to ask for peace, which on our part was as willingly granted them. Many speeches were made on the occasion, and, if it may be allowed to judge from gesture and the language of nature, many of them possess qualifications that would be no disgrace to a modern orator. Although there is much repetition

[12] Identified July 25, 1902, by R. Kent Beattie, Principal Pathologist, Bureau of Plant Industry, U. S. Department of Agriculture, Washington, D. C. (Letter, Jan. 30, 1937, Harold St. John, Professor of Botany, University of Hawaii, Honolulu, to the author; letter, June 9, 1937, R. Kent Beattie to the author). The water is now used for irrigation. A description of the spring is given by Israel Cook Russell in "Geology and Water Resources of Nez Perce County, Idaho," *Water Supply and Irrigation Papers of the U. S. Geological Survey*, Washington, D. C., no. 53 (1901), pp. 79-81.

in their harangues, delivered with much vehemence and intense feeling, they are uniformly natural and are certainly calculated either to tie the knot of affection and sympathy, or rouse the mind to discord and war. I have observed speakers and hearers so overcome that they sobbed and cried aloud, and the proceedings delayed until they recovered. This affair was concluded in the usual way—exchange of presents.[13]

Leaving the site of Lewiston and traveling overland with Work's party and their purchases, 114 fine horses, Douglas arrived at Fort Colville on August 4, after an absence of two months. At Spokane House, where he stopped on the way, he had a hearty welcome from old Jocko, this time with food—fine fresh salmon, now plentiful, as many as two thousand being caught in a single day by the Indians with bone-pointed spears. During the journey several species of evening primrose (*Oenothera*), clover (*Trifolium*), sagebrush (*Artemisia*), and a novel *Eriogonum* were added to his collection.

At Fort Colville he devoted ten days to collecting seeds of the more important plants seen there earlier in the summer, getting additional plants, and drying and packing. Having heard from McLoughlin that a ship was to leave the Columbia for England early in September, he determined that he must reach the coast somehow in order to forward his collections by it. First, however, in accordance with his plan for returning to England via Hudson Bay next spring, he packed a small box with a portion of his seeds and paper and two shirts, to be sent across the Rocky Mountains to Fort Edmonton, where he would get it on his way.

The Columbia was at its most dangerous season, high and turbulent with floodwater from its glacier-fed tributaries, and its many rapids, cascades, and whirlpools rendered a small canoe unsafe. A larger one was not available, so Chief Trader Dease at Fort Colville arranged that the first lap of the journey

[13] Douglas, *Journal*, pp. 201-202. For John Work's account of this quarrel, see *Washington Historical Quarterly*, VI (1915), 34-35.

DOUGLAS FIR TREES

FORT VANCOUVER

JASPER HOUSE

toward the seacoast should be on horseback, with the Little Wolf, a chief of the Okanagan tribe, as guide. But the Little Wolf had more serious business on hand. An old quarrel had just broken out between his people and a party of the Kootenay Indians from Columbia Lake, the headwater of the river. Both parties were now engaged in the salmon fishing at Kettle Falls, where tempers had reached a white heat. The opposing warriors met with strung bows and loaded guns, stark naked except for their war-caps of calumet-eagle feathers and the paint on their bodies. As the affair threatened to involve the whites, Dease stepped forward at a critical moment and knocked down a warrior who was aiming his arrow at a chief of the other tribe, probably saving the chief's life. He urged them to go home peaceably. The whole day was spent in clamor and haranguing; the Little Wolf, being one of the principals in the affair, could not go that day. The morrow brought more uproar, concluding toward evening with peace, sealed by an exchange of presents. This too required the attendance of the Little Wolf, who was "equally wanted by his party, whether to make war or peace." Then a great feast was held to celebrate the event; and since it was uncertain when the Little Wolf would be free from his duties, Douglas obtained another guide, Robado, of the Spokane Tribe.

Taking three horses and supplies of pemmican, sugar, and tea, and clad in deerskin trousers presented by Dease, he left Fort Colville on August 19 for Fort Okanagan. He took a tent but was unable to pitch it owing to the barren country and lack of wood. For the same reason campfires had to be made from grass and from dried horse-droppings found at old encampments. The alkali-laden water in the small lakes was unfit to drink, so he suffered much from thirst, as did the horses. The extreme heat necessitated starting each day at two o'clock in the morning, with a few hours rest at noon if the mosquitoes permitted it. Another difficulty was his inability to speak a word of the guide's language, so that signs had to be used. Despite these obstacles they made good time, and arrived at Fort Okanagan on the 22nd.

Quickly procuring two Indians and a canoe, Douglas proceeded downstream, himself acting as steersman. He arrived at Fort Walla Walla on August 25 too weak to eat, owing to the exhausting travel from daylight to dark and the loss of his tea and sugar through a swamping of the canoe. A tin shaving pot which served as his only cooking utensil was similarly lost. An Indian who guided them through a difficult passage "thought himself well paid with a few crumbs of tobacco and a smoke out of my own pipe." In spite of his exhaustion, next day with two fresh guides and a larger canoe he hurried on, traveling all night since he knew the bad places in the river and wished to pass a camp of notorious Indian pillagers while they slept.

He changed canoes again at the Great Falls. At the Dalles he was wrecked and lost all the insects collected on his journey to the interior, as well as some seeds and his pistol. At the portage he found several hundred Indians gathered. Chief Pawquanawaka was away, but the "Chiefess" entertained him with nuts and whortleberries. Here Douglas felt in his "own province" again, and as he understood the language he had no difficulty in obtaining a small canoe and two guides, one, called Red Coat, being well known to him.

The natives welcomed him back from up-country and in the evening a party of no less than seventy-three came to smoke with him. Things went pleasantly until he missed his tobacco box. "As soon as I discovered my loss I perched myself on a rock, and, in their own tongue, gave the Indians a furious reprimand, applying to them all the epithets of abuse which I had often heard them bestow on another; and reminding them that though they saw me only a *Blanket Man,* I was more than that, I was *the Grass Man,* and therefore not at all afraid of them. I could not, however, recover my box, but slept unmolested after all the bustle." [14]

[14] Douglas, "Sketch," *OHQ,* V, 368.

At the Cascades the water was so rough and the wind so high that he had to carry all his belongings on his back along the tall shelving rocks, while the Indians brought down the canoe. He was welcomed by Chief Chumtalia (his quasi-guide of last year), who treated him to a hearty meal of salmon and whortleberries served on a mat. The chief's brother and nephew accompanied him to Fort Vancouver, where he arrived on August 30, having traveled the six hundred miles from Fort Colville[15] in twelve days accompanied only by Indians.

He had successfully navigated the great but treacherous waterway of northwest America—the river of which another traveler has said: "Many lives are lost in this river. None but Canadians and Indians would ever think of navigating this terrible Columbia." [16]

As he plodded wearily up from the landing place in his scanty and worn-out clothing the men of the Fort were alarmed. They had never seen the river higher, and feared he was the sole survivor of a drowning accident. Their frightened glances astonished him, but after quickly dispelling their fears he received a genuinely hearty welcome. Among the new faces to greet him was Peter Skene Ogden, noted fur trader, whom he found "a man of much information and seemingly a very friendly-disposed person." Known as "Monsieur Pete" to the voyageurs, Ogden came of a well-known Montreal family and had been educated for the law; but a high-pitched, squeaky voice and a roving temperament had turned him from pleading to fur trading.[17]

[15] The distance by water all the way (Symons, *Report on Upper Columbia River*, pp. 52-54). The distance traveled was probably less, since the first part of the journey was by land and no doubt shorter than by river. Although Douglas gave the distance as nearly 800 miles (*Journal*, p. 211), elsewhere he gave it as 620 (130 plus 490) miles (*ibid.*, p. 255).

[16] Rev. H. H. Spalding, in *Oregon Historical Quarterly*, XIII (1912), 375.

[17] Although favorably disposed toward Douglas, Ogden regarded later visiting naturalists as a nuisance: "Last summer . . . by the Snake country we had . . . five more Gent. as follows 2 in quest of Flowers 2 killing all the Birds in the Columbia & 1 in quest of rocks and stones all these bucks came with letters from the President of the U. States and you know it would not be good policy

Douglas's race to Fort Vancouver was successful; he arrived just in time, for the *Dryad,* Captain Davidson, was to sail on the second day after his arrival. Only a day was left in which to unpack and repack in chests his collections of the year, including one hundred twenty species of seeds, and get them sent out to the ship. Several letters, too, had to be written, taking far into the night. But fatigued as he was, he had the satisfaction of knowing that the hard-won gleanings of another season and of a new and extensive territory were on their way to England. During the past five months the dry, sandy regions of the interior had felt a botanist's footstep for the first time, just as the densely wooded regions of the lower Columbia had the year before.

Relaxing after his arduous travels, he leisurely gleaned a few seeds of choice plants collected last year: red-flowering currant, salal, broad-leafed maple, vine maple, and tall Oregon grape; "laid in specimens of *Pinus taxifolia* [Douglas fir] with fine cones; and collected a few sections of the various woods, gums, and barks of the different timbers that compose the forest" in the neighborhood of Fort Vancouver.

---

not to treat them politely they are a perfect nuisance" (letter, Feb. 25, 1837, Ogden to John McLeod, in *Washington Historical Quarterly,* vol. II, 1908, p. 260).

9

IN SEARCH OF THE SUGAR PINE

THE comparatively comfortable quarters of the gentlemen of the Hudson's Bay Company at Fort Vancouver did not harbor Douglas long. He had not recovered from his exhausting up-river itinerary before another journey was planned. A year ago on his journey up the Willamette he had found the Indians eating like nuts the seeds of a strange pine which they said grew in the mountains farther south. Since they had no perfect specimens of seed he had asked Baptiste McKay,[1] who was going in that direction, to obtain for him a dozen cones, some twigs and gum, and a small bag of seeds. McKay had returned to Fort Vancouver in February with nothing except one cone—all he could obtain. It was enough, however, to further excite Douglas's admiration, and led him to write enthusiastically to Hooker about the as yet unseen pine:

> I rejoice to tell you of a new species of *Pinus*, the most princely of the genus, perhaps even the grandest specimen of vegetation. It attains the enormous height of from one hundred and seventy to two hundred and twenty feet, with a circumference of fifty feet, and cones from twelve to eighteen inches long! I possess one of the latter measuring one foot five inches long and ten inches round the thickest part. The trunk grows remarkably straight and destitute of branches till near the top, where they form a perfect umbel; the wood of fine quality, yielding a large quantity of resin. Growing trees of this *Pinus*, which have been partly burnt by the natives to save themselves the trouble of collecting fuel, a custom to which they are greatly addicted, produce a substance which, I am almost afraid to

[1] Jean Baptiste Desportes McKay (often referred to as Desportes McKay), an Astorian overlander of 1809 (*Washington Historical Quarterly*, vol. XXIV, 1933, pp. 288-290).

say, *is sugar;* but as some of it, together with the cones, will soon reach England, its real nature will then be correctly ascertained. This *Pinus* is found abundantly two degrees south of the Columbia River, in the country of the Umptqua tribe of Indians, who collect its seeds in autumn and pound them into a kind of cake, which they consider as a kind of luxury, using also the saccharine substance that I have described above, in the same way as civilized nations do sugar. I intend to bring home such an assemblage of specimens as will allow a correct figure to be taken of this tree, and also to try my success with a bag of its seeds.[2]

He had also sent an account of it to Sabine,[3] naming it *Pinus Lambertiana* after Lambert, author of the great work on the genus *Pinus*.[4] It is a curious fact that, although Douglas's name was given to the former *Pinus taxifolia*, the Douglas fir, it was the sugar pine which particularly aroused his curiosity and led him to make one of the most daring expeditions of his career.

After his return to Fort Vancouver from the interior he again turned his attention to the new pine. He had intended accompanying McKay on his annual fur hunt to the south, but found he had gone, having left some weeks earlier than usual. However, on consulting McLoughlin he learned that a party under Chief Trader Alexander Roderick McLeod was to visit the same quarter shortly, and he arranged to accompany them. As the country was new (in fact, totally unknown south of the Umpqua River) and the journey had to be made mostly by land, his outfit was confined to his gun, a small tent, two blankets, two spare shirts, tobacco for presents and to pay for his return journey, a small copper kettle, six quires of plant paper, and incidentals.

[2] Douglas, "Sketch," *OHQ*, V, 330.

[3] *Transactions*, Linnean Society, London, XV (1827), 497-500.

[4] The common name, sugar pine, is derived "from the sugary exudation, sought by the native tribes, which forms hard, white, crystalline nodules on the upper side of fire or axe wounds in the wood. This flow contains little resin, is manna-like, has cathartic properties, and is as sweet as cane sugar, but belonging to a different class of sugars, namely pinite or pine sugar" (Willis Linn Jepson, *The Silva of California*, Memoirs of the University of California, Berkeley, Cal., vol. II, 1910, p. 72). The seeds, too, are edible, and were valued by the natives for their flavor (*ibid.*).

He left Fort Vancouver on September 20 by boat with a party under Donald Manson and reached McLeod's camp fifty-six miles up the Willamette[5] two days later. Thence the journey was on horseback in a southerly direction up the west side of the river. Douglas was fortunate in having one of McLoughlin's best horses to ride or to carry his belongings, as he chose. The party (now under McLeod and numbering thirty) made slow progress, averaging only a few miles a day because of a variety of difficulties: first, the frequent straying of the horses from the camping places, sometimes requiring a day's delay to find them; second, the hunting of deer for meat (they were to live off the country), Douglas taking his turn with the others at hunting; third, the poor trail, necessitating the cutting away of branches and fallen trees, the fording of streams, and the ascending and descending of steep hills, which hurt the horses' feet; and fourth, the rainy weather.

Douglas walked most of the way in order to botanize, but his feet got sore from the wiry grasses and the burnt stumps of low brushwood. Elsewhere thickets of brake ferns and wild blackberries made walking very fatiguing. Moreover the territory along the Willamette was not good for botanizing, for a good deal of it had been burnt off.[6] According to some of the Indians this had been done to compel the deer to feed in the unburnt areas, where they could be readily detected and killed. Others said it was done "in order that they might the better find wild honey and grasshoppers, which both serve as articles of winter food."[7] Douglas found some new plants, however, including

[5] Afterwards the starting point for the California Brigade, and called "Campement du Sable" or "Champoeg" (George B. Roberts, "Recollections of the Hudson's Bay Company," MS., Bancroft Library, Berkeley, California, p. 100).

[6] Douglas appears to have been the first to mention forest fires in the Pacific Northwest.

[7] The grasshoppers, like the deer, would be driven to the unburnt areas, where they could be caught in large numbers. At the same time the fires would make way for a lush growth of fresh plants (including the well-known fireweed) upon which deer, grasshoppers, and bees would thrive. This seems to be the first record of honey in the Oregon territory. Whether it came from wild bees or from bumble bees is not stated.

two specimens of *Rosa,* and a new *Ribes.* On the banks of the Santiam River (near where the town of Jefferson stands) he discovered the now familiar, strong-smelling camphor weed (*Trichostem lanceolateum*), which was used by the Indians to stupefy fish so that they could be easily obtained.[8] He also gave his attention to the various birds and animals: deer, elk, squirrels, partridges, describing them in detail in his journal, and obtaining specimens of such as he could carry.

The dangers and hardships of the journey soon began. One of the party, John Kennedy, while out hunting was attacked by a large male grizzly bear and barely escaped with his life by climbing a tree, after leaving most of his clothing with the bear. The shortage of food was often distressing, living from hand to mouth as they were; but what they did get they appreciated. Douglas relates an occasion when "McKay made us some fine steaks, and roasted a shoulder of the doe for breakfast, with an infusion of *Mentha borealis*[9] sweetened with a small portion of sugar. The meal laid on the clean mossy foliage of *Gaultheria Shallon*[10] in lieu of a plate and our tea in a large wooden dish hewn out of the solid, and supping it with spoons made from the horns of the mountain sheep or Mouton Gris of the voyageurs."[11]

The grizzlies had their compensating features; when McKay shot a large female, Douglas traded an old blanket and a little tobacco for the skin to use as an under-robe for sleeping on the damp, cold ground. Travel continued on Sunday as on other days. The men observed it by changing their linen, and in the evening those who could read perused tracts consonant with Roman Catholic doctrine.

The farther they went the more difficult became the route.

[8] Haskin, "David Douglas."
[9] The Canada mint.
[10] The salal.
[11] Douglas, *Journal,* p. 221.

Leaving the Willamette valley, they crossed over three ranges of mountains before entering the valleys of Pass Creek and Elk Creek[12] (the latter, called Red Deer River by Douglas,[13] a tributary of the Umpqua).[14] McLeod and Douglas took the lead in finding the way through the thickly wooded country and in hewing down branches that obstructed the horses. Rain had made the footing so bad that several of the animals fell on the steep hills and rolled down until arrested by trees, stumps, or brushwood.[15] Fearing an accident to his collections, Douglas wrapped them up in a bearskin and carried them on his back, adding to them an odd deer horn and other interesting natural history specimens as he went along.

Here the country was more favorable for botanizing. The prodigious size of the numerous fallen trees (probably *P. taxifolia*) amazed him, some measuring two hundred forty feet long and eight feet in diameter. He was greatly pleased with a new tree found growing on the hills, the Western Chinaquin, or golden leafed chestnut (*Castanopsis chrysophylla*),[16] pronouncing it "a princely tree. . . . Nothing can exceed the magnificence of this tree, or the strikingly beautiful contrast formed with the sable glory of the shadowy pine among which it delights to grow."

In the valley of Elk Creek he found another delightful tree, the California laurel (*Umbellularia californica*), a splendid evergreen distinguished from all others by the strong, camphoric-pungent odor of its crushed leaves or green bark, which,

[12] Albert R. Sweetser, in *Sunday Oregonian*, Portland, Nov. 7, 1926, sec. 5, p. 5.

[13] Called La Biche by McLeod (Maurice S. Sullivan, *The Travels of Jedidiah Smith*, Santa Ana, Cal., 1934, pp. 120, 179).

[14] The trail afterwards opened to California began at the junction of the Elk and the Umpqua Rivers (Gustavas Hines, *Wild Life in Oregon*, New York, 1887, pp. 97, 98).

[15] The trail was still bad in 1840 (*ibid.*, pp. 98, 99).

[16] George B. Sudworth, *Forest Trees of the Pacific Slope* (Washington, D. C.: United States Department of Agriculture, Forest Service, 1908), pp. 273-276.

when rubbed in his hands, made him sneeze.[17] He learned that a concoction made from its bark was used by the hunters as a drink, and declared it "by no means an unpalatable beverage." He hoped these trees were new discoveries, having had in the case of the former a laborious search up a steep hill to find one in fruit. The latter he hoped would be useful in medicine or as a perfume, besides being a valuable addition to the Society's garden. In order to obtain seeds he was obliged to cut down a large tree, the bark being too smooth to climb. The chestnut did turn out to be new,[18] but the laurel had been found by Menzies.[19] Several species of *Clethra* were gathered; and, among birds, some specimens of condor (*Sarcoramphos californica*) and quail (*Ortyx californica*) were obtained.

On October 16 the Umpqua (or Arguilar) River was reached at the junction of Elk Creek (afterwards the site of Fort Umpqua,[20] where Elkton now is). Douglas described the place as "bed sandstone; ninety yards broad; not deep, but full of holes and deep chinks worn out by the water. Two hundred yards below is a small rapid in several channels and small grassy islands; will never admit of any barge larger than a ship's jolly-boat by the numerous rocks that rise above water and the

[17] "The green bark and, particularly, the leaves possess a light volatile oil, follicles of which are given off when either is crushed, and which when inhaled through the nostrils produces severe pain over the eyes, attended often by violent sneezing" (Sudworth, *Forest Trees,* p. 327). Douglas was unable to introduce either of these trees. Regarding the former, see F. R. S. Balfour, "David Douglas," *Journal of the Royal Horticultural Society,* LXVII (1942), 126-128.

[18] W. J. Hooker, *Flora Boreali-Americana* (London, 1840), II, 159; Jepson, *Silva,* p. 241. Douglas apparently did not take seeds home. See also *Curtis's Botanical Magazine,* London, 1856, plate 4953.

[19] W. J. Hooker, *Flora Boreali-Americana,* II, 137; *Curtis's Botanical Magazine* (1862), plate 5320; but even earlier the Spanish Californians knew the tree, calling it *Laurel silvestre* (Jepson, *Silva,* p. 245). Apparently the seeds that Douglas collected were the first to reach England (*Curtis's Botanical Magazine, supra*).

[20] Bancroft, *Northwest Coast,* II, 521.

rapids" [21]—a description that still holds.[22] It was a territory little known, seldom visited by the fur traders.[23]

They arrived exhausted from the trail-cutting and the shortage of food, and, finding nothing to shoot, they were supperless. To add to the misery, some of the horses, Douglas's among them, were too worn out and too far behind to bring in that night, so that Douglas was bedless as well as supperless. Consequently he and McLeod took turns at sitting by the fire and sleeping in the latter's blanket and buffalo robe. In the morning they were fortunate in shooting a 190-pound deer.

Several days previously Baptiste McKay had sent one of his hunters to obtain some cones of Douglas's "much-wished-for Pine," but he returned with only a few baked seeds. Finding among the Umpqua Indians two who spoke Chinook, Douglas questioned them about the possibility of reaching the tree. Reassured by their answers, he started off up the Umpqua with one of McKay's Indians, intending to reach the mountains to the southeast. (His route was similar to that of the present road between Elkton and Oakland.) [24] After going nine miles it was necessary to cross the river, so he undertook to make a raft. He made it too small, however, and blistered his hands so badly that he sent the Indian guide back to McLeod's camp for help. While the guide was gone Douglas went hunting, and in pursuing a large buck which he had shot through the shoulder he fell into a deep gully among some deadwood, where he lay unconscious for five hours until he was extricated by some Calapooie Indians. As his chest had been hurt and was very painful, he gave up the journey and began to return to McLeod's camp,

[21] Douglas, *Journal,* p. 223.
[22] Letter, Feb. 7, 1938, A. R. McDonald, Postmaster, Elkton, Oregon, to the author. Douglas's camp is said to have been on the opposite side of the river from the fort side (Sweetser, in *Sunday Oregonian,* Portland, April 12, 1925, sec. 5, p. 11). The "Old Fort" (Umpqua) was near the junction of Calapooya Creek and the Umpqua (Sullivan, *Travels,* pp. 113, 178).
[23] Barnston, "Abridged Sketch," p. 202.
[24] Sweetser, in *Sunday Oregonian,* Portland, Nov. 7, 1926, sec. 5, p. 5.

his horse with his belongings being led by an Indian while he crept along with the help of a stick and his gun. Part way he was met by John Kennedy, whom McLeod had dispatched to assist with the raft, and was given his horse to ride to camp. On arriving he relieved his pain by bleeding himself in the left foot, and next morning bathed in the cold river, after which rigorous treatment he felt much better.

During the next few days he accompanied the McLeod party down the Umpqua river. A large black-tailed deer was shot and proved an object of great interest to him. Likewise of interest was the Indians' snare for catching deer, made of a small species of Iris "which, though no thicker than the little finger, is strong enough to secure the largest Buffalo and the Elk."

The natives, knowing little of white people, were of a suspicious attitude, but on learning of the peaceful intentions of the party they condescended to supply them with food and canoes. The faces of nearly all the women were tattooed, which was done with a sharp piece of bone and a cinder from the fire. "It is considered a great mark of beauty," wrote Douglas. "I have little doubt that such a lady in London would make a fine figure, particularly when a little red and green earth is added to the upper part of the face."

But the "much-wished-for Pine" had yet to be found. Opportunity for another effort came when the party arrived at tidewater, twenty-three miles from the sea. They were visited by Centrenose, the principal chief from the upper country, and McLeod arranged with him for his son to accompany Douglas there on another search for the sugar pine. So on October 23 he retraced his steps up the river, while McLeod's fur-hunting party proceeded to the sea and southerly along the coast. He took only two horses, one for the guide and one for his own things, he himself walking.

On reaching the place of his unsuccessful raft-making he obtained passage across the river in a canoe. At the home of the the guide's father he was hospitably entertained with salmon trout, hazel nuts, roasted nuts of the California laurel, camas roots, and meal made from a syngenesious plant. Douglas reciprocated with presents of deer flesh, beads, rings, and tobacco. "Among the riches of his father's home were fifteen wives, one of whom he was at pains to make me understand was his mother. He had himself been wedded only a few days, and had some reluctance of leaving his young bride."

They proceeded in an easterly direction, hampered by bad weather and shortage of food—enough for only one meal a day. Fortunately the journey did not take many days. The story of the discovery of the sugar pine is told by Douglas as follows:

Wednesday 25th.—Last night was one of the most dreadful I ever witnessed. The rain, driven by the violence of the wind, rendered it impossible for me to keep any fire, and to add misery to my affliction my tent was blown down at midnight, when I lay among *Pteris aquilina*[25] rolled in my wet blanket and tent till morning. Sleep of course was not to be had, every ten or fifteen minutes immense trees falling producing a crash as if the earth was cleaving asunder, which with the thunder peal on peal before the echo of the former died away, and the lightning in zigzag and forked flashes, had on my mind a sensation more than I can ever give vent to; and more so, when I think of the place and my circumstances. My poor horses were unable to endure the violence of the storm without craving of me protection, which they did by hanging their heads over me and neighing. Towards day it moderated and before sunrise clear, but very cold. I could not stir before making a fire and drying part of my clothing, everything being completely drenched, and indulging myself with a fume of tobacco being the only thing I could afford. Started at ten o'clock, still shivering with cold, although I rubbed myself with my handkerchief before the fire until I was no longer able to endure the pain. Shortly after I was seized with a severe headache and pain in the stomach, with giddiness and dimness of sight; having no medicine except a few grains of calomel, all others being done, I could not think of taking that and therefore threw myself into a violent perspiration and in the evening felt a little relieved. Went through an open hilly country some thirteen miles,

[25] *Pteridium aquilinum* (Brake fern).

where I crossed the river to the south side[26] near three lodges of Indians, who gave me some salmon such as is caught in the Columbia and at this season scarcely eatable, but I was thankful to obtain it. Made a short stay and took my course southerly towards a ridge of mountains, where I hoped to find my pine. The night being dry I camped early in the afternoon, in order to dry the remaining part of my clothing. Travelled eighteen miles.

Thursday, 26th.—Weather dull and cloudy. When my people in England are made acquainted with my travels, they may perhaps think I have told them nothing but my miseries. That may be very correct, but I now know that such objects as I am in quest of are not obtained without a share of labour, anxiety of mind, and sometimes risk of personal safety. I left my camp this morning at daylight on an excursion, leaving my guide to take care of the camp and horses until my return in the evening, when I found everything as I wished; in the interval he had dried my wet paper as I desired him. About an hour's walk from my camp I was met by an Indian, who on discovering me strung his bow and placed on his left arm a sleeve of racoon-skin and stood ready on the defence. As I was well convinced this was prompted through fear, he never before having seen such a being, I laid my gun at my feet on the ground and waved my hand for him to come to me, which he did with great caution. I made him place his bow and quiver beside my gun, and then struck a light and gave him to smoke and a few beads. With a pencil I made a rough sketch of the cone and pine I wanted and showed him it, when he instantly pointed to the hills about fifteen or twenty miles to the south. As I wanted to go in that direction, he seemingly with much good-will went with me. At midday I reached my long-wished *Pinus* (called by the Umpqua tribe *Nàtele*), and lost no time in examining and endeavouring to collect specimens and seeds. New or strange things seldom fail to make great impressions, and often at first we are liable to over-rate them; and lest I should never see my friends to tell them verbally of this most beautiful and immensely large tree, I now state the dimensions of the largest one I could find that was blown down by the wind: Three feet from the ground, 57 feet 9 inches in circumference; 134 feet from the ground, 17 feet 5 inches; extreme length, 215 feet. The trees are remarkably straight; bark uncommonly smooth for such large timber, of a whitish or light

[26] Near the present Cleveland, Oregon, post office (Sweetser, in *Sunday Oregonian,* Nov. 7, 1926, sec. 5, p. 5).

brown colour, and yields a great quantity of gum of a bright amber colour. The large trees are destitute of branches, generally for two-thirds the length of the tree; branches pendulous, and the cones hanging from their points like small sugar-loaves in a grocer's shop, it being only on the very largest trees that cones are seen, and the putting myself in possession of three cones (all I could) nearly brought my life to an end. Being unable to climb or hew down any, I took my gun and was busy clipping them from the branches with ball when eight Indians came at the report of my gun. They were all painted with red earth, armed with bows, arrows, spears of bone, and flint knives, and seemed to be anything but friendly. I endeavoured to explain to them what I wanted and they seemed satisfied and sat down to smoke, but had no sooner done so than I perceived one string his bow and another sharpen his flint knife with a pair of wooden pincers and hang it on the wrist of the right hand, which gave me ample testimony of their inclination. To save myself I could not do by flight, and without any hesitation I went backwards six paces and cocked my gun, and then pulled from my belt one of my pistols, which I held in my left hand. I was determined to fight for life. As I as much as possible endeavoured to preserve my coolness and perhaps did so, I stood eight or ten minutes looking at them and they at me without a word passing, till one at last, who seemed to be the leader, made a sign for tobacco, which I said they should get on condition of going and fetching me some cones. They went, and as soon as out of sight I picked up my three cones and a few twigs, and made a quick retreat to my camp, which I gained at dusk. The Indian who undertook to be my last guide I sent off, lest he should betray me. Wood of the pine fine, and very heavy; leaves short, in five, with a very short sheath bright green; cones, one 14½ inches long, one 14, and one 13½, and all containing fine seed. A little before this the cones are gathered by the Indians, roasted on the embers, quartered, and the seeds shaken out, which are then dried before the fire and pounded into a sort of flour, and sometimes eaten round [whole]. How irksome a night is to such a one as me under my circumstances! Cannot speak a word to my guide, not a book to read, constantly in expectation of an attack, and the position I am now in is lying on the grass with my gun beside me, writing by the light of my Columbian candle—namely, a piece of wood containing rosin.[27]

[27] Douglas, *Journal*, pp. 229-231.

The place where Douglas had these thrilling experiences was probably what is now known as Sugar Pine Mountain, west of Roseburg, Oregon.[28]

Next morning he was visited by three grizzly bears. He shot two, and paid his guide with the carcass of one of them. He then returned down the Umpqua. At one place his horse nearly fell into the river, being saved by becoming wedged fast between two fallen trees. Douglas tied its legs and head to prevent it from struggling, then freed it by cutting away the lower tree.[29] He "felt a great deal on this occasion, as the horse had been Mr. McLoughlin's present to me, and was his own favourite animal."

Three days' travel brought him back to his starting place. Here he found Michel Laframboise and an Indian boy, who reported the Indians very troublesome since the departure of the hunting brigade. There was no sleep that night. A large fire was made; they hid in the grass a little way off to watch, wet through from the rain; and toward morning they drove off a party of fifteen Indians by firing blank shot.

Next night a similar watch was kept but there were no marauders. The following day Baptiste McKay returned from the seacoast and they felt much relieved; he was not only a bulwark against treacherous natives but also an expert hunter. Next morning he shot a large doe long-tailed deer, which was served with a big kettle of rice soup made by Douglas, making a fine supper for the small party as well as for thirteen fur hunters just back from the coast. Altogether it was a pleasant evening and Douglas found it "pass away agreeably to the eleven preceding, and although the society at many times uncouth, yet to have a visage of one's own colour is pleasing; each gave an account of the chase in turn."

McLeod returned a day or two later, having been down the

[28] Sweetser, in *Sunday Oregonian*, April 12, 1925, sec. 5, p. 11; John D. Guthrie, in *Oregonian*, Portland, Ore., July 14, 1932.

[29] Rev. Gustavas Hines had a similar experience here in 1840 (Hines, *Wild Life in Oregon*, pp. 115-116).

coast as far as the Coquille River. Since two of his men were being dispatched to Fort Vancouver, Douglas decided to return with them. He much appreciated McLeod's constant assistance during the search for the sugar pine, and made a note in his journal that on getting back to London he must "get him a good rifle gun as a present." McLeod likewise formed a good opinion of the botanist: "A most devoted and indefatigable man, and . . . always the best of company in a journey." [30]

The party, consisting of Douglas, John Kennedy (an Irishman), and Fannaux (a Canadian), with nine horses, started out on November 7. They experienced an exhausting two weeks' journey, owing to the usual shortage of food for themselves as well as for the horses, the bad weather, and the difficulty of finding the trail. To drink they had soup made of pounded camas roots, which made Douglas sick. He longed for "a little tea, the greatest and best of comforts after hard labour." Again one meal a day was the rule, and he records that it was not good for either tempers or travel:

> *Saturday, 11th.*—Last night, after lying down to sleep, we began to dispute about the road, I affirming we were two or three miles off our way, they that we were quite close to our former encampment; all tenacious of our opinions. The fact plainly this: all hungry and no means of cooking a little of our stock; travelled thirty-three miles, drenched and bleached with rain and sleet, chilled with a piercing north wind; and then to finish the day experienced the cooling, comfortless consolation of lying down wet without supper or fire. On such occasions I am very liable to become fretful.[31]

In the morning he obstinately went off alone and got lost, necessitating Kennedy's coming back to find him. However, he made up for it by shooting some geese and ducks which relieved the hunger of the party. The horses became worn out, and for a time it was feared that some of them would have to be abandoned, but by resting them they were saved. There was con-

[30] Markham, *Paxton and the Bachelor Duke*, p. 70.
[31] Douglas, *Journal*, pp. 235-236.

tinual hardship, but despite it all there was little grumbling. Douglas observes that "there is a curious feeling among voyageurs. One who complains of hunger or indeed of hardship of any description, things that in any other country would be termed extreme misery, is hooted and brow-beaten by the whole party as a *pork-eater* or a young voyageur, as they term it."

His worst experience came near the end of the journey, when in crossing the Santiam River (they had to swim) the greater part of the collection that he had gathered at the cost of weeks of tireless effort and severe hardship was lost. Fortunately the precious pine cones and some other specimens were saved.

The party arrived at Fort Vancouver on November 19. Two months' hard travel through a new and difficult country had been undertaken by Douglas in search of one tree; but it had been found. Notwithstanding the loss of most of his collection, and in spite of his being laid up for nine days on his return with painful swollen ankles caused by continued exposure to wet and cold, it had been a successful journey. Its main purpose had been accomplished.

And the tree for which he had so exerted himself was indeed worthy of his efforts. Its massive, straight, grayish-brown trunk and high-up, horizontal branches of deep blue-green foliage tipped with clusters of long cones make it a magnificent tree, while its soft, light wood is of high commercial value. A leading Californian botanist has declared that "the Sugar Pine is undoubtedly the most remarkable of all pines, viewed either from the standpoint of its economic value or sylvan interest." [32]

One would have thought that after his arduous and harrowing experiences of the spring, summer, and autumn, Douglas would have spent the winter in rest and recuperation. But no; that was too much for his restless spirit. He scarcely had recovered from the effects of his last journey when, finding "time lying heavy"

[32] Jepson, *Silva*, p. 73. Although it had a wider meaning in Douglas's day, and sometimes has still in popular parlance, the term "pine" as used here and by modern botanists in general does not include the Douglas fir, which of course excels the sugar pine in grandeur and value.

on his hands, he visited the seacoast in search of seaweeds, shells, "or anything that might present itself." The weather was so bad that his canoe was dashed to pieces on the shore. On arriving at his friend Chief Cockqua's he was obliged to take pot-luck with him, dried salmon and salal berries being the only fare he could offer. This gave Douglas an attack of diarrhea lasting several days, which so weakened him that he was unable to walk. On his way back, nearly starved, he had to trade the handkerchief off his neck and seven buttons off his coat for two bits of fish. He managed to get to Fort Vancouver on Christmas Day after a sixteen days' journey as fruitless as that to the same quarter a year before.

Another journey to the coast in March was also unproductive. The interval was spent at Fort Vancouver in arranging the plants collected, drying and putting up seeds, collecting specimens of forest timber and cryptogamic[33] plants, and copying some notes regarding the Chinook tribe.

The long winter evenings would have been intolerably dull had it not been for the companionship of the other sojourners at the Fort. Among those who arrived by the overland express in November was Chief Trader James McMillan, who in the following summer was to found Fort Langley on the Fraser River.[34] Then there was George Barnston, who afterwards was in charge at Fort Walla Walla. Natural history was his hobby and in it he received much encouragement from his contacts with Douglas. After his retirement from the fur trade he was for a time president of the Natural History Society at Montreal. There was also Lieutenant Aemilius Simpson, R.N., who had been sent from London to superintend the Hudson's Bay Company's shipping on the Pacific and captain the *Cadboro,* first vessel to enter the Fraser River. He is remembered for the casual way in which apple trees were introduced. At a farewell dinner upon his leaving England a lady friend, more in jest

[33] Producing neither flower nor seed, such as ferns, mosses, algae, etc.
[34] Bancroft, *Northwest Coast,* II, 477-484.

than in earnest, dropped into his waistcoat pocket some apple seeds for him to plant at his wilderness destination. He thought no more about them until after his arrival, when he wore the waistcoat again and found them still in the pocket. From those seeds were grown the first apple trees in the Pacific Northwest.[35]

The winter of 1826-27 was long and cold, with hard frosts and much snow; and spring was late in coming. The confinement indoors was irksome to Douglas, but in spite of it he was regarded as "one of the heartiest, happiest mortals" in the little society at the lonely outpost of empire.[36]

Barnston records that the weather was so bad that many of the horses died. This caused various beasts and birds of prey to hover about, most conspicuous among which was the California condor. "This magnate of the air," he says, "was ever hovering around, wheeling in successive circles for a time, then changing the wing as if wishing to describe the figure 8; the ends of the pinions, when near enough to be seen, having a bend waving upwards, all his movements, whether of soaring or floating ascending or descending, were lines of beauty. In flight he is the most majestic bird I have seen. One morning a large specimen was brought into our square, and we had all a hearty laugh at the eagerness with which the Botanist pounced upon it. In a very short time he had it almost in his embraces fathoming its stretch of wings, which not being able to compass, a measure was brought, and he found it full nine feet from tip to tip. This satisfied him, and the bird was carefully transferred to his studio for the purpose of being stuffed. In all that pertained to nature and science he was a perfect enthusiast." [37]

[35] *Ibid.*, II, 441

[36] Barnston, "Abridged Sketch," p. 207.

[37] *Ibid.*, p. 208.

10

HOMEWARD BY HUDSON'S BAY EXPRESS

Douglas had now spent nearly two years in the Columbia River region, a term of successful but hard work, dangerous and exhausting journeys, and privations of all sorts. His time to leave had come, so, torn between the pleasure of turning homeward and the regret at leaving "a country so exceedingly interesting," where he had spent "if not many comfortable days, many pleasant ones," he left Fort Vancouver with the annual express on March 20, 1827. His final tasks had been packing four boxes of his collections to send home by ship, and making a tin box for his journals and as many of his seeds as he could expect to take with him to Hudson Bay.

Other passengers with the express were Dr. McLoughlin and Messrs. A. R. McLeod, Annance, and Pambrun, all Company men destined to up-river points. There were two boats: the express boat, and another which was to assist at the Great Falls portage and strengthen the party against the Indians, then return to Fort Vancouver.[1] In command was Edward Ermatinger, just turned thirty, with the rank of clerk.

The journey as far as Fort Colville was not new to Douglas, nor was he unaccustomed to the express; he had traveled with it last year as far as the Spokane River, walking most of the

[1] The boats used on the Columbia were of different construction from those used east of the Rocky Mountains. They were flat-bottomed, tapering toward the ends, and propelled by paddles instead of oars. The crew had consisted of eight men until reduced to seven by Governor Simpson on his visit to the Columbia in 1824-25. The crews of the two boats were required to transport one boat at a time over the portages (Frederick Merk, *Fur Trade and Empire; George Simpson's Journal: . . . 1824-1825,* Cambridge, Mass., 1931, p. 38).

way.[2] Now, again, he was on foot a great deal ("the whole distance," he declares in one journal),[3] partly because the boat was overcrowded, and partly in order to collect botanical, zoological, and mineralogical specimens. Although he had been over the ground before, "not a day passed but brought something new or interesting either in botany or zoology." A few mineral specimens also were found. But his discoveries were made at the expense of his feet, which became "very painful, blistered, and blood-run."

At Fort Colville, where they arrived on April 12, he envied several fine animal skins which John Work had procured at the request of Deputy Governor Nicholas Garry in London: a pair of white mountain sheep (*Mouton blanche*), a pair of black-tailed deer, and a small wolf of "a singular variety and curious from its being the deity or god of the Flathead tribe of Indians." He was consoled, however, by the thought that ultimately they would probably reach his friend Sabine of the Horticultural Society. He dug up some bulbs of spring beauty and mission bell, and some roots of dogtooth violet, which were so attractive that, although it was a bad time for removal, he tried to transplant them all the way to England.

He was greatly upset when some famished dogs belonging to the Indians seized and devoured a pair of grouse which he had shot on the way and was taking to England. "Although they were closely tied in a small oilcloth and hung from the tent-poles, the dogs gnawed and ate the casing, which were leather thongs. Grieved at this beyond measure. Carried the cock bird 457, and the hen 304 miles on my back, and then unfortunately lost them."

Bidding farewell to his kind Columbian friends, he left Colville on the 17th, his company now being confined to Ermatinger and the boat crew of four Canadians and three Iroquois Indians. He found the young clerk a most agreeable companion

[2] See Chapter 7.
[3] Douglas, *Journal,* p. 69.

and an accomplished player on the flute, with which he entertained the party at the evening campfires.

The journey on the Upper Columbia had much of interest for Douglas. His keen eye and busy pencil noted the course of the river, its width, its current, its tributaries, the slope of its shores, the vegetation and timber, and the mountains towering over all. He attempted a polished description but felt it inadequate:

> "Sky beautiful at sunset, the snowy summits of the hills tinted with gold; the parts secluded from his rays are clothed with cloudy branches of the pine wearing a darker hue, while the river at the base is stealing silently along in silvery brightness or dashes through the dark recesses of a rocky Dalle. How glad should I feel if I could do justice to my pencil (when you get home, begin to learn)." [4]

When the river widened out to form a lake (now Lower Arrow Lake) and a forward breeze sprang up, a sail was hoisted, enabling the voyageurs to rest from their labor and engage in some rollicking boat songs. Douglas, however, was annoyed by the music; just then he was trying to write down some words of the Chinook dialect. But it was a good day for traveling: they made about forty-seven miles.

Continuing on through the upper lake (now Upper Arrow Lake)[5] and along the river, they came to a dangerous whirlpool and rapid (now Big Eddy and Steamboat Rapid, just beyond the little city of Revelstoke). "The most terrific grandeur" of the scene impressed Douglas, the narrowly confined river, the "rapids, whirlpools, and still basins, the water of a deep dark hue, except when agitated. On both sides high hills with rugged rocks covered with dead trees, the roots of which being laid bare by the torrents are blown down by the wind, bringing with them blocks of granite attached to their roots in large masses, spreading devastation before them. Passing this place just as

[4] Douglas, *Journal*, p. 248.

[5] The Upper and Lower Arrow Lakes got their names from the Indian arrows which had been shot into fissures in the steep rocky shore of the lower lake and become lodged there.

the sun was tipping the mountains and his feeble rays now and then seen through the shady forests, imparts a melancholy sensation of no ordinary description, filling the mind with awe on beholding this picture of gloomy wildness." [6] To get the boat up the rapid required the combined strength of the entire party, two in the boat guiding it with poles, and seven pulling it by line from the shore. Douglas carried all his belongings on his back "lest evil should overtake them."

The next three days' travel was through a series of dangerous rapids. At the worst of them, the Dalles des Morts (Death Rapids), back-packing was again in order; only after three hours' labor and anxiety was the boat lined up and the baggage carried around what have been called "those dangerous whirlpools, whose huge serpentine throats were darkened the more by the grand overhanging forests and perpendicular cliffs of the closed up mountains through which the Columbia, God knows when, wore down her way." [7] This was one of the most dangerous places on the river, where many lives had been lost. Here a rough wooden cross, there a solitary grave, told their silent story.[8]

[6] Douglas, *Journal*, pp. 251-252.

[7] Angus McDonald, "A Few Items of the West," p. 215.

[8] Alexander Ross, *The Fur Hunters of the Far West* (London, 1855), II, 179. The ominous name originated in 1817, when seven men were wrecked and all their food was lost. They began walking along the river hoping to reach Spokane House, the nearest establishment, over three hundred miles away. The high water forced them up into the almost impenetrable forest. One by one they died, the survivors resorting to cannibalism. The last one was found by Indians on the shore of the upper (Arrow) lake and was taken to Kettle Falls, whence he was conducted to Spokane House. His story that he had killed his last companion in self defense was not believed, and he was dismissed from the North West Company's service, escaping more serious punishment owing to the lack of evidence against him. Varying accounts of the disaster, some with gruesome details, are given by Ross Cox, *Adventures on the Columbia River* (London, 1831), II, 181-184; Paul Kane, *Wanderings of An Artist among the Indians of North America* (London, 1859), 329-331; Douglas, *Journal*, pp. 252-253. On October 22, 1838, a boat of the westbound express capsized at the first rapid below the Dalles des Morts and drowned several persons, including two young English gardeners, Robert Wallace and Peter Banks, who had been sent out by Sir Joseph Paxton on a botanical expedition to follow up Douglas's work (Markham, *Paxton and the Bachelor Duke*, pp. 63-72).

Gladly leaving the Dalles des Morts behind them, Ermatinger's party continued on to the head of the Big Bend of the Columbia. Here Douglas got his first view of the Rocky Mountains: "Although I have been travelling for the last fifteen days surrounded by high snowy mountains,[9] and the eye has become familiar to them and apt to lose that exalted idea of their magnitude, yet on beholding those mentioned impresses on the mind a feeling beyond what I can express. I would say a feeling of horror."

The journey up the Columbia was ended; it was now April 27—a month and a week since leaving Fort Vancouver. The sturdy boat was cached on the shore at Boat Encampment[10] to await the coming of the westbound express in the autumn, when it would be used for going down the river.

As his travels on the Columbia were over, Douglas made a list of them in his journal, showing a total of 7032 miles, as follows:

In 1825

	Miles
From the ocean to Fort Vancouver, on my arrival in April	90
In May, to and from the ocean to Fort Vancouver	180
In June, to and from the Great Falls	210
In July, to and from the ocean and along the coast	216
In August, journey on the Multnomah River	133
In September, to the Grand Rapids	96
On the mountains of the Grand Rapids	47
In October and November, to the sea	90
In the same, trip to Cheecheeler River or Whitbey Harbour	53
Ascending said river	65
Portage from it to the Cowalidsk River	35
Descending the latter	40
An allowance of my daily wanderings from Fort Vancouver, my headquarters	850
	2105

[9] The Selkirk Mountains.

[10] Situated at the mouth of Wood River on its left bank (*Annual Report of Lands and Survey Branches of Department of Lands,* British Columbia, for 1936, Victoria, B. C., 1937, p. O 35)

In 1826

In March and April, from Fort Vancouver to the Kettle Falls	620
In May, journey to Spokane	150
In June, from Kettle Falls to the junction of Lewis and Clarke's River	414
In June, journey to the Blue Mountains	190
In July, a second to the same	137
In July, ascending Lewis and Clarke's River to the north and south branch	140
A third journey to the Blue Mountains from that place	103
From Lewis and Clarke's River to Spokane	165
From Spokane to Kettle Falls	75
In August, from Kettle Falls to Okanagan by land	130
From Okanagan to Fort Vancouver	490
In September, October, and November, from the Columbia to the Umpqua River and the country contiguous thereto	593
To the ocean and the bays north of the Columbia in December	125
Daily allowance from my places of rendezvous	600
	3932

In 1827

In March and April, the whole chain of the Columbia from the ocean to the Rocky Mountains	995
Total	7032

Apart altogether from their botanical results these journeys in themselves were a great achievement. With the exception of Lewis and Clark and the fur traders, no white man had previously made any of them; and nobody—not even a fur trader—had undergone them all.

The next stage of the express was to be on foot: the crossing of the Rocky Mountains, via Athabaska Pass. The lofty snow-covered peaks which had given Douglas such "a feeling of horror" when viewed from a distance were to be encountered at close quarters. To prepare for the long tramp he examined and repacked his tin box of seeds and wrapped in a handkerchief the clothing he was not wearing. For walking on the snow he had four pairs of snowshoes of the bear-paw type (that is, rounded at both ends), one having been presented by John Work at Fort Colville, and the other three having been purchased from Indians on the way up the river. Ermatinger kindly proposed that all the botanist's belongings be carried by others of the party so that he could travel unencumbered; but, seeing Ermatinger's load, Douglas declined, and insisted that the least he could do was to carry the seeds and journals which already had cost him such labor and anxiety. Accordingly, they were wrapped in oilcloth and strapped on his back. They weighed forty-three pounds.

Leaving Boat Encampment on April 28, the party began its journey up the Wood River valley to Jeffrey Creek,[11] up the gully of that stream and over a high, timbered ridge[12] to a rivulet flowing from Athabaska Pass, and so to the Pass.[13] East of the summit the route was down the Whirlpool River to the Athabaska.[14]

The first day's travel was through woods and swamps along Wood River. In crossing a swamp three miles long the men frequently sank to their knees, becoming doubly fatigued by having to break through the thin ice on the surface. In the woods the snow was from four to seven feet deep. Douglas, being a novice with snowshoes, found them very awkward, and had several falls head over heels. He was bothered by the lacings' getting wet and coming loose and by the snowshoes' turning backside foremost when entangled in the brushwood.

Nevertheless he studied and noted the vegetation, animals and birds along the way and collected some willow specimens. Even the arduous travel he found interesting: "To-day is a scene

[11] Stream names given subsequent to Douglas's journey.

[12] Called "Grand Cote" by Ermatinger, and "Big Hill" by Douglas.

[13] *Report of the Inter-provincial (Alberta-British Columbia) Boundary Survey Commission,* part II (Ottawa, 1924), p. 88; *Canadian Alpine Journal,* Banff, XII (1921-22), 166-167.

[14] *Report of the Inter-provincial Boundary Survey Commission,* II, 92.

of some curiosity even to myself, and I can hardly imagine what a stranger would think to see nine men, each with his load on his back (food and clothing), his snowshoes in his hand, starting on a journey over such an inhospitable country: one falling, a second helping him up, a third lagging and far behind, a fourth resting smoking his pipe, and so on."[15]

The next day was spent walking along the gravel bars of the river and snowshoeing through the woods. They forded the river fourteen times, plunging up to the waist in the nearly freezing water. It was not only piercing cold, but dangerous; the current was so strong that if it had gotten under the soles of their feet it would have swept the men away, so they kept their feet on the bottom and dragged them across; in some places the entire party had to hold onto one another and cross obliquely to avoid disaster. Sodden and shivering after each crossing, they hurriedly put on their snowshoes and hastened onward to warm and partially dry themselves before the next plunge.

"Found the cold piercing," says Douglas, "alternately plunging to the middle in water 35° Fahrenheit and skipping with my load to recover my heat among the hoar frost. At 9 A.M. entered a point of wood where the snow was 4 to 7 feet deep, with a weak crust not strong enough to support us. Obliged to put on my bears' paws; path rough, and in addition to the slender crust, which gives the traveller more labour, were dead trees and brushwood lying in all directions, among which I was frequently caught."[16]

Such were the experiences of transcontinental travelers by the Hudson's Bay "express," be they Company officers, employees, or "passengers."[17]

[15] Douglas, *Journal*, p. 256.

[16] Douglas, *Journal*, p. 257.

[17] Eastbound in 1825, Governor Simpson's party forded the river sixty-two times in one day, supporting one another to avoid being swept away (Ross, *Fur Hunters*, II, 192). Westbound in the spring of 1846, Father de Smet had an even worse experience, crossing the swollen "Great Portage River [now Wood River] not less than forty times, with the water frequently up to our shoulders." The party was obliged to support one another in crossing, to avoid

Little time or strength was left for anything else. Douglas shot a wolverine but it got away. He tried to catch a wood partridge alive but had to shoot it. Although they had started at four in the morning the day's progress was only fifteen miles. That evening he "made a pair of socks out of the legs of a pair of old stockings; the feet being worn, took the skirts of my coat to wrap round my toes instead of socks. Strict economy here is requisite: my feet, ankles and toes very painful from the lacing of my snowshoes; otherwise, well and comfortable, lying in a deep hole or pit among the snow on a couch of pine branches with a good fire at my feet. If good weather visits us, we are thankful; if bad, we make the best of a bad situation by creeping each under his blanket, and, when wet, dry it at the fire." [18]

The following day after eleven more fordings of the river they came to the Big Hill. Here, owing to the thick woods, the steep ascent, and the deep snow, Douglas had his troubles:

The ravines or gullies unmeasurable, and towards noon becoming soft, sinking, ascending two steps and sometimes sliding back three, the snowshoes twisting and throwing the weary traveller down (and I speak as I feel) so feeble that lie I must among the snow, like a broken-down waggon-horse entangled in his harnessing, weltering to rescue myself. Obliged to camp at noon, two miles up the hill, all being weary. No water; melted snow, which makes good tea; find no fault with the food, glad of anything. The remainder of the day is spent as follows: On arriving at a camp, one gathers a few dry twigs and makes fire, two or three procuring fuel for the night, and as many more gathering green soft branches of *Pinus balsamea*[19] or *canadensis*[20] to sleep on, termed 'flooring the house,' each hanging up his wet clothing to the fire, repairing snowshoes, and arranging

being carried away by the impetuous current. The continual soaking swelled his legs, all his toenails came off, and his moccasins were stained with the blood (Pierre J. de Smet, *Oregon Missions and Travels over the Rocky Mountains in 1845-46*, New York, 1847, p. 209).

[18] Douglas, *Journal*, p. 257.

[19] *Abies balsamea*, Mast. in *Journal of the Royal Horticultural Society*, XIV, 189. (This note and the next are editor's notes from Douglas, *Journal*, p. 258.)

[20] *Tsuga canadensis*, Mast., *ibid.*, p. 255.

his load for the ensuing day, that no time may be lost; in the morning, rise, shake the blanket, tie it on the top, and then try who is to be at the next stage first. Dreamed last night of being in Regent Street, London! Yet far distant. Progress nine miles.[21]

Next morning the campfire, having been kindled on the snow, had sunk into a hole six feet deep, "making a natural kitchen." A good fire had been needed, for although it was the first of May the temperature was down to two degrees above zero. After a laborious tramp up the Big Hill, frequently losing the trail because blazes on the trees were hidden by the deep snow, they plodded on and eventually came to Athabaska Pass.

While the group rested here Douglas indulged in the mountaineering exploit which is the subject of Chapter 11.

To add to the party's meagre food supply Ermatinger shot a species of bird, the flesh of which was prized by the voyageurs for its tenderness. None of them got even a taste of it, however, for when Douglas saw it he had other plans. It was the first of its kind he had seen, and he was so struck with its beauty that he "could not resist the temptation of preserving it" and adding it to his load. He took it with him all the way to England, where in reward for his perseverance he had the pleasure of reading a paper before the Linnean Society of London in which he described it in detail and named it *Tetrao Franklinii* (Franklin's Grouse), in honor of Captain John Franklin, the Artic explorer.[22] It eventually found its way to the Royal Scottish Museum at Edinburgh, where it still may be seen, so well did Douglas preserve it in bleak Athabaska Pass more than a century ago.

The following morning they came to the summit of the Pass and "one of the head springs of the Columbia, a small lake or basin twenty yards in diameter, circular, which divides its waters, half flowing to the Pacific and half to the hyperborean sea—namely, the headwaters of the Athabasca River."

This was the Committee's Punch Bowl (so named by Gover-

[21] Douglas, *Journal*, p. 258.
[22] *Transactions*, Linnean Society, XVI (1833), 139.

nor George Simpson in honor of the Company's committee of management),[23] whose oddity of situation and outflow, and distinctive name, were to bring it such notoriety that it was shown prominently on some of the maps.[24] Douglas and Ermatinger seem not to have heard of the name yet, for their journals do not mention it.[25]

Here in Athabaska Pass, at an elevation which he greatly overestimated as 12,000 feet above sea level, Douglas discovered the elegant little species of the primrose tribe afterwards named after him, the *Douglasia nivalis* or *Snow Douglasia.*[26] The only bird that he noticed in the vicinity was "a small light dun Jay, who, with all the impudence peculiar to most of his kindred, fluttered round our camp last night picking any food thrown to him"—a fairly good description of the now well-known whisky jack or camp robber.

Although the Committee's Punch Bowl was considered the halfway point in the journey, the more difficult and dangerous part was over. Descending the Pass northeasterly along the Whirlpool River (formerly called Punch Bowl Creek)[27] the party gladly increased its pace. The descent was steeper than the ascent had been, the weather was milder, there was less snow than on the west side of the mountains, and the herbage

[23] Merk, *Fur Trade,* p. 34; Ross, *Fur Hunters,* II, 198.

[24] Map accompanying "Report on a Journey in British Columbia," by Lt. Richard C. Mayne, in *Journal of the Royal Geographic Society,* London, XXXI (1861), 213; map in Mayne's *Four Years in British Columbia and Vancouver Island* (London, 1862); map in Capt. John Palliser's *Journals, Detailed Reports, and Observations* (London, 1863). Bancroft, in his *History of British Columbia,* p. 37, calls it the "most remarkable of anything of the kind on the planet."

[25] The name is mentioned in revisions of Douglas's journal written after his arrival in England. In one of these he coldly but truly remarks: "This is not the only fact of two opposite streams flowing from the same lake." The name was given by Simpson in 1824.

[26] John Lindley, in *Edwards's Botanical Register,* XXII (1836), table 1886, and *Quarterly Journal of Science,* London, II (1827), 383-385; *Paxton's Magazine of Botany,* London, III (1837), 236. The height of Athabaska Pass is 5724 feet. Although the plant may have been found on the mountain slope above the Pass it could not have been at anything like 12,000 feet. Further, see Chapter 11.

[27] Ross, *Fur Hunters,* II, 200.

was different. The giant cedars and firs of the Columbia gave way to the small black spruce and jack pine. "The difference of climate and soil," said Douglas, "with the amazing disparity in the variety and stature of the vegetation, is truly astonishing; one would suppose it was another hemisphere, the change is so sudden and so great." Three prominent mountains and a large glacier were passed (now mounts Kane, Evans, and Scott, and Scott Glacier,[28] in Jasper National Park).

The first person they met on the east side of the mountains was Jacques Cardinal, pioneer packer and guide in the Jasper region, who had brought horses up the trail for them. Douglas stayed at his camp for the night and found him a generous host, although his means of entertainment were primitive:

Old Cardinal roasted, on a stick before the fire, a shoulder of *Mouton gris,* which I found very fine. He had a pint copper kettle patched in an ingenious manner, in which he was boiling a little for himself; this with a knife was all the cooking-utensils. He observed he had no spirit to give me, but turning round and pointing to the river he said 'This is my barrel and it is always running.' So having nothing to drink out of, I had to take my shoulder of mountain sheep and move to the brook, helping myself as I found it necessary.[29]

The party proceeded on horseback (Douglas preferring to walk) to the end of the Rocky Mountain portage at the confluence of the Miette and Athabaska Rivers (opposite the present Jasper village). Embarking there in canoes, they reached Jasper House,[30] "three small hovels," on May 4. Their overnight stay was celebrated by a stag dance, the music supplied by Ermatinger's servant on an old violin found at the post.

[28] Lawrence J. Burpee, *On the Old Athabaska Trail* (London, 1927), pp. 132-133.

[29] Douglas, *Journal,* pp. 260-261.

[30] Jasper House then was situated at the outlet of Brule Lake, about thirty-five miles northerly from the present village of Jasper. In 1830 it was moved to the outlet of Jasper Lake (letter, Sept. 18, 1935, J. Chadwick Brooks, Secretary, Hudson's Bay Co., to the author, quoted in A. G. Harvey, "The Mystery of Mount Robson," *British Columbia Historical Quarterly,* Victoria, B. C., vol. I, 1937, pp. 222-223).

Mount Brown *The Committee's Punch Bowl (uppermost lake)*

ATHABASKA PASS

MOUNT HOOD

The day after leaving Jasper House they overtook George McDougall and a party of four men en route from Fort St. James on Stuart Lake in New Caledonia to Fort Carlton on the Saskatchewan River, where they were to obtain a supply of moose or caribou hides for use as leather in making moccasins, bags, clothing, etc. This was one of the first of the leather brigades which annually used this route for many years.[31] McDougall's party had experienced a terrible journey up the Fraser and had run short of food. When one of the men had been caught stealing the pemmican, he had been summarily shot and killed by McDougall to protect the lives of the others—a striking instance of the stern justice meted out by the officers of the Hudson's Bay Company.[32]

Ermatinger's party traveled the 184 miles between Jasper House and Fort Assiniboine in three long days, the journey being made more difficult by the river ice which necessitated portages in several places. Douglas found this part of the journey very monotonous: "Seeing one mile gives an idea of the whole." There was nothing striking in the way of plants, but he noted the many horizontal beds of coal along the river's bank.

The express route left the Athabaska at Fort Assiniboine and went overland to Fort Edmonton. On arriving at Fort Assiniboine, however, Ermatinger found that scarcity of provisions obliged him to wait for a supply which was being brought by Chief Factor John Stuart and party from Lesser Slave Lake. To assist Stuart in ascending the river a party went down from Fort Assiniboine, and Douglas accompanied them to see the country. The two parties met at the junction of the Lesser Slave River. Douglas found that Stuart had a more intimate knowledge of the great Northwest than anyone he had encountered, besides being familiar with plants and other departments of natural

[31] A. G. Morice, *History of the Northern Interior of British Columbia* (Toronto, 3rd ed., 1905), pp. 157-158.

[32] T. C. Young, "Extracts from George McDougall's Journal," MS., British Columbia Archives, Victoria. Morice, *History*, p. 159, gives extracts from McDougall's journal but does not mention the homicide.

history. In 1808 he had accompanied Simon Fraser on the memorable descent of the river bearing his name; afterwards he was on the Columbia. Douglas had been wondering whether to sail for England from Montreal or New York instead of from Hudson Bay, but upon Stuart's recommendation he adhered to his decision in favor of the latter, being advised that the longer canoe trip to either of the eastern ports would give little opportunity for natural history collecting, would be expensive, and would cut short by six weeks his time in the West.

Learning that Finan McDonald, to whom he had entrusted his seeds last year for carriage across the continent, had fallen in with Thomas Drummond, assistant naturalist under Captain John Franklin, while descending the Athabaska, a note of professional jealousy crept into Douglas's journal: "Hope my box is safe (do not relish botanist coming in contact with another's gleanings)."

Five days of "cheerless and comfortless" travel had been taken in meeting Stuart's party and returning to Fort Assiniboine. Leaving there for Fort Edmonton on May 14 he found the country "uninterestingly wretched," since it afforded him no plants. Stuart had provided him with two horses, one to carry his collection, "and one to ride, but being averse to that he [the horse] went light." Consequently Douglas walked the entire distance (about a hundred miles), floundering up to the waist through bogs, swamps, and mire, crossing the Paddle, Pembina, and Sturgeon Rivers on hastily constructed rafts, and crossing the smaller streams on trees hurriedly felled. Provisions were so scanty the express had to slow up while the men caught fish to enable it to proceed, and the horses broke down from hunger.

Anxious to have a little time for collecting at Fort Edmonton and to know the fate of his box of seeds, Douglas hurried ahead of the brigade, accompanied by an old Nipissing Indian, and in one day walked the remaining forty-three miles. He thus narrates his arrival on the evening of May 21:

> Night creeping in on me, my view of the country gradually disappeared. At eight, on reaching a rising eminence unexpectedly, I heard the evening howl of the sledge-dogs, which to me was sweet music, and perceived fires in some lodges which I knew to be near the establishment. Being all over with mud, I returned half a mile to a small lake, stripped and plunged myself in and then comforted myself with a clean shirt which I carried on my back in a bundle. I was most kindly received by Mr. John Rowand, and had supper prepared for me of fine moose-deer steaks, which were most acceptable after a walk of forty-three miles through a most wretched country without having anything to eat.[33]

Too tired to sleep, he arose at daybreak to examine his precious seeds. He found them in good order, considering the trouble McDonald had had in carrying them across the mountains, only eighteen papers having been damaged by moisture. However, among these he was sorry to find his *Paeonia*, "one of the finest plants in the collection. It often happens that the best goes first." Further damage had been prevented by his shirts' absorbing some of the moisture. This he considered fortunate, although their loss (having become entirely rotten) was a serious matter. Greater damage would have resulted had not Drummond, at Chief Factor Rowand's suggestion, changed the papers while visiting the Fort, for which kind act Douglas's jealousy changed to gratitude.

He found little of botanical interest at Edmonton owing to the lateness of the season. However, he did obtain an interesting bird: a fine young specimen of the noble and rare golden eagle, which was presented to him by Rowand. It was known as the calumet eagle in the Northwest because its tail feathers were highly prized by the Indians for decorating their ceremonial pipes, or calumets, as well as for adorning their war caps and other garments. Douglas heard surprising stories of its strength and ferocity, such as its carrying off young deer alive and killing full-grown long-tailed deer; and he noticed that on its approach all other birds left their prey in the utmost terror.

[33] Douglas, *Journal*, p. 267.

His present had been brought all the way from the Kootenay district. It had become quite a favorite with the boys of the Fort, who had teased it and pulled out much of its plumage, putting it in bad temper which caused Douglas some difficulty in caging it.

The evening before his departure was made merry with a dance in his honor, the music being supplied by Ermatinger on a violin. When Douglas "was given to understand it was principally on my account, I could not do less than endeavour to please by jumping, for dance I could not."

The express left Fort Edmonton on May 26. The Stone Indians were hostile, making it unsafe to sleep at a campfire, so after traveling all day the party drifted downstream all night with their boats lashed together, two by two. (They were now descending the North Saskatchewan River.) Douglas found "this mode of travelling very irksome" and "ill adapted for botanizing," since they were "never on shore except a short time when cooking breakfast, always dusk before a second meal." Nevertheless he managed to add to his collection specimens of milk vetch, phlox and loments, and seven species of willow. His opportunities for collecting were increased by several stops which the party made for hunting buffalo.

On one of these hunts a deadful accident befell his friend of the seed box, Finan McDonald, who was terribly gored by a wounded buffalo bull and left for dead. Douglas, the only man of science available, ministered to him, finding him unconscious but alive, with a punctured chest, two broken ribs, a thigh laid open to the bone, and other injuries. The lancet which the botanist always carried in his pocket for use on animals and birds was put to use in bleeding McDonald, notwithstanding that blood had already been running from his wounds for two and a half hours—harsh treatment even for a hard-bitten fur trader; but such was the faith in the practice of bleeding. His wounds were bound up and he was carried to the river and placed in a boat, where Douglas gave him a stiff dose of laudanum (twenty-five drops) to put him to sleep. He was then hurried downstream some three hundred miles to Cumberland House, a five and a half days' journey, in order to have the services of Dr. John Richardson of Franklin's party.

Whether because of Douglas's field surgery or Richardson's professional attendances or his own good constitution (he was a man of enormous stature and wonderful physique), McDonald recovered, although his death was reported.[34] Indeed, he later became a Member of Parliament in Upper Canada and injected into his forensic battles some of the fighting tactics he had used in the West against Indians and buffaloes.[35] It is said that he was the original of the character MacDonald Bhain in Ralph Connor's book, *The Man from Glengarry.*[36]

At Carlton House, Douglas met Thomas Drummond, and at Cumberland House, Dr. John Richardson, naturalists with the second land Arctic expedition under Captain John Franklin, now returning home. Both had made extensive collections of plants, birds, and quadrupeds.[37] After examining them, Douglas declared each one a "princely collection." They lessened his regret at his own limited opportunities for botanizing on the present journey.[38]

Reaching Lake Winnipeg by way of Cedar Lake and the Grand Rapids, the express crossed the northern end of Lake Winnipeg to its outlet, and descended about twenty miles to Norway House, the old crossroads of the North, at its new establishment on Playgreen Lake. Here the express arrived

[34] Dr. John Richardson, *Fauna Boreali-Americana* (London, 1829-1831), I, 281.

[35] J. A. Meyers, "Finan McDonald—Explorer, Fur Trader and Legislator," *Washington Historical Quarterly*, XIII (1922), 196.

[36] E. O. S. Scholefield and F. W. Howay, *British Columbia* (Vancouver and Chicago, 1914), III, 838.

[37] John Franklin, *Narrative of a Second Arctic Expedition to the Shores of the Polar Sea, in the years 1825, 1826, and 1827* (London, 1828); Thomas Drummond, "Sketch of a Journey to the Rocky Mountains and to the Columbia River in North America," in W. J. Hooker's *Botanical Miscellany*, London, I (1830), 178-219.

[38] Letter, July 9, 1827, Douglas to Clinton, Columbia University.

on June 16, and Douglas was pleased to find letters from England. One from Sabine of the Horticultural Society brought good news: the safe arrival of the plants sent by ship from the Columbia, the fullest satisfaction with his work, and complete approval of all that he had "done, written and projected," including his delayed return to England.[39] It also brought melancholy news: the death of Douglas's father. This was confirmed by a letter from his brother.

Next morning Governor Simpson arrived en route from Montreal to Hudson Bay.[40] His Excellency was interested to hear of Douglas's travels and collections, and, noticing that he was poorly clad, offered him a change of linen and other clothing. For some reason—perhaps modesty—Douglas declined it with thanks.

Since the date of his ship's departure for England allowed time for a side trip to the Red River Colony (320 miles), Douglas left the express at Norway House.[41] While waiting for a boat he went over the plants collected during his descent of the Saskatchewan River, tabulating them and changing the papers. In spite of the difficulties of collecting, he had gotten no less than ninety-four varieties.

When Captain Franklin arrived at Norway House and offered passage up Lake Winnipeg in his canoe, Douglas accompanied him, leaving on July 2. He botanized at the camping places along the lake shore, until they arrived at Fort Alexander at the mouth of the Winnipeg River on July 8. Here Franklin left him and proceeded to Montreal.[42]

Obtaining a canoe and a guide, Douglas went on around the

[39] Letter, March 10, 1827, Joseph Sabine to Douglas, copy at Royal Horticultural Society. The Horticultural Society gave it out that Douglas "has been induced to remain another season," and that "his expedition . . . has hitherto succeeded beyond the most sanguine expectation of his employers." *Transactions*, VI (1826), iv-v.

[40] Franklin, *Narrative*, p. 314.

[41] He had intended going from Carlton House across country to Red River but gave up the idea owing to the Stone Indians' being hostile.

[42] Franklin, *Narrative*, p. 314.

south shore of the lake and up the Red River, doing his share of the paddling and putting his hands "in sheets of blisters." He continued his botanizing, collecting seventy-four varieties. His approach to the colony and his welcome he describes as follows:

> At sunrise passed several thinly planted low houses, with small herds of cattle wandering from the folds; humble and peasant-like as these may appear to many, to me—who have been no sharer of civilized society for a considerable time past—they impart a pleasant sensation. At seven took breakfast two miles below the rapid, where I left the canoe and my luggage to go by land, taking with me my boards and paper. Strangers in this quarter appear to be few: scarcely a house I passed without an invitation to enter, more particularly from the Scottish settlers, who no doubt judging from my coat (being clothed in the Stewart or royal tartan) imagined me a son from the bleak dreary mountains of Scotland, and I had many questions put to me regarding the country which now they only see through ideal recollection.[43]

Upon arriving at Fort Garry on July 12 he was received by Donald McKenzie, governor of the colony, who entertained him during his stay. McKenzie had an intimate knowledge of the West and the Far West, and had traveled so much that he was called "Perpetual Motion" McKenzie.[44]

A visitor from the outside world to the little settlement (the beginning of the present city of Winnipeg) was an event; and Douglas soon had callers. The first was a pupil at the school of the Church Missionary Society, an Indian boy from the Columbia named Spokane Garry, who came to inquire about his father and brothers whom Douglas had met. He and another boy were the first natives of the Columbia district to be taught to read and write.[45] Their subsequent reports to their people regarding Christianity are said to have led to the Flathead delegation's

[43] Douglas, *Journal,* p. 279.

[44] Ross, *Fur Hunters,* I, 283. A recent biography glorifies him as "King of the Northwest" (Cecil W. Mackenzie, *Donald Mackenzie, "King of the Northwest,"* Los Angeles, Cal., 1937). The surname has had various spellings.

[45] Ross, *Fur Hunters,* II, 156-158.

going to St. Louis for Christian teachers, which resulted in the sending of the first missionaries to the Indians of Oregon.[46]

Other callers were Bishop J. N. Provencher, Rev. Theophilus Harper, and Mr. Buchier of the Roman Catholic Mission, who made inquiries regarding the territories Douglas had visited. He felt quite honored, for their call was the first they had made on anyone other than a Hudson's Bay Company official. A few days later he dined with the bishop and spent the evening in looking over the garden, farm, church, and school where the Indian children were taught not only religion, but also domestic economy, farming, spinning, and weaving cloth from the wool of the buffalo.

He called on Rev. David T. Jones, the Church of England missionary and principal of the Protestant school, and on his assistant, Rev. William Cochran. He attended either the Protestant or the Catholic Church every Sunday. After his two years of wandering through the Far West in the company largely of Indians and voyageurs he enjoyed the companionship of these educated men, regardless of their faiths. Reverend Harper accompanied him on a two-day botanizing trip to White Horse Plain, eighteen miles up the Assiniboine River, where they were guests of Cuthbert Grant, noted chief of the half-breeds and Warden of the Plains.

He spent nearly a month at the Red River Colony, botanizing diligently and collecting 110 varieties. On August 10 he left for Hudson Bay, accompanied by Dr. Richard J. Hamlyn, the Company and Colony surgeon. The voyage down Lake Winnipeg was very stormy, with almost hurricane winds and enormous waves which harassed them whether afloat or ashore. Near the foot of the lake Douglas was pleased at finding specimens of *Linnaea borealis* (twinflower; a favorite plant of Linnaeus) in fruit, all his efforts to find it in that state on the Columbia having been unsuccessful.

[46] Clifford M. Drury, "Oregon Indians in the Red River School," *Pacific Historical Review*, Glendale, Cal., VII (1938), 50-60.

The route from Norway House to Hudson Bay was down East River to Echimamish River, up that river, and across Painted Stone Portage to Whitewater Lake, a headwater of the Hayes River. This stream was descended to Hudson Bay, passing through Oxford and Knee Lakes on the way. At a bad rapid about halfway down the river the boat smashed against a rock, and Douglas and his companions just managed to reach a small island as the craft filled. It was so tricky a place that the voyageurs sometimes undressed before reaching it so that they might more easily swim if suddenly immersed.[47] They reached York Factory on August 20, where to Douglas the view of the Hudson's Bay Company's ship riding at anchor was inspiriting. The *Prince of Wales* meant home!

He received a warm welcome from Chief Factor John George McTavish, and found waiting for him a new suit of clothes and other necessary garments. Probably these had been provided by direction of Governor Simpson, who, in spite of the refusal of his offer two months before, realized Douglas's predicament and the propriety of his arriving in England decently clad. This offer Douglas did not refuse.

He was delighted to find that the bulbs and roots which he had brought all the way from Fort Colville were alive and fresh; and in a note in his journal he reproached himself for not having tried the same with his favorite plant, the lowly but sprightly salal: "Why did you not bring *Gaultheria* alive—across the continent—2900 miles? It could be done."

In the case of the birds that he did bring a good part of the way, however, he had bad luck. When he arrived at Norway House from the West he had with him two live eagles which he hoped to get safely to England. One, a silver-headed or bald eagle, he left at Norway House during his trip to Red River. The other, the golden eagle presented by Rowand at Fort Edmonton, he sent on to Hudson Bay because of the shortage of meat at Norway House and the eagle's dislike for fish. Great was

[47] Garry, "Diary," pp. 149-150. He had a "miraculous escape" there.

his disappointment on his return from Red River to learn that the silver-headed eagle had died of starvation. More bitter still was the blow on reaching York Factory, when he learned that a few days previously the golden eagle had caught in its leash and been strangled to death. "What can give one more pain?" he wrote. "This animal I carried 2000 miles and now lost him, I might say, at home."

Despite its disappointments, he might well have been proud of the hard and dangerous journey from the Pacific. It was the longest he had made in America, and, added to the previous ones, plus side trips to the Lesser Slave and Red rivers, made a grand total of about ten thousand miles traveled by canoe, row-boat, or on foot, horseback, or snowshoe during a little over two years. It was another great achievement. Again it can be said that, excepting the fur traders, he was the first person to make the journey. Although the northerly way taken by the Hudson's Bay express was the earliest regularly used route across the continent, never before had it been traveled from end to end by an outsider.[48]

Those years were a noteworthy period for scientific travel in the fur country, for besides the explorations of Douglas there were those of Franklin, Richardson, and Drummond. Eleven years were to pass, however, before the express journey was again attempted by anyone in the interest of science, and then it was cut short by disaster (the drowning of Wallace and Banks, already mentioned).

A last incident remained—one which nearly brought Douglas's career to an end in the icy waters of Hudson Bay. One day while awaiting the departure of the ship a party consisting of himself and Drummond, Back, and Kendall of Franklin's expedition, with eight oarsmen, went out to visit it. (It was anchored some distance from shore owing to shallow water.) On their

[48] Established in 1814 by the North West Company, it originally continued on from Norway House to Montreal. Following the union of the two companies the eastern terminus was changed to York Factory.

way back in the evening the wind turned into a hurricane. At the mercy of wind and waves, the boat was tossed about like an eggshell and driven out to sea. Rowing was out of the question, and all that the men could do was to bail out some of the water with their hats. Night settled down on them accompanied by tremendous thunder and rain, and the great rolling billows con tinually broke over them. They became so numb with cold that it was difficult to bail out fast enough to keep the boat from going under. All of the party suffered greatly and Douglas became quite ill.

The storm continued all night, and although it lessened the next afternoon it was not until the middle of the following night that oars could be used. They were now out of sight of land, with no compass, and sun invisible; as it turned out, they were about seventy miles out in the bay. However, by strenuous efforts and with the aid of the tide they made their way back to the ship and then to the shore, to find that they had been given up for lost. Douglas and Kendall suffered severely from the experience and did not recover the full use of their limbs until after arriving in England.[49]

The *Prince of Wales* sailed from York Factory on September 15 and arrived at Portsmouth on October 11. Although it was an excellent passage,[50] Douglas was too ill to enjoy it, being confined to bed most of the time.[51] Even his journal-writing ceased, no mention being made in it of the voyage nor of the terrible experience in the bay.

[49] Drummond, "Sketch," pp. 216-218.
[50] *Ibid.*
[51] W. J. Hooker, "Memoir," *OHQ*, VI, 218.

11

PUTTING MOUNTAINS ON THE MAP

THE READER may have memories of school days when he was taught that of all the mountains in North America the highest were Mount Brown with 16,000 feet and Mount Hooker with 15,700. For the greater part of a century these names and challenging figures stood out on the best official maps, and it is only within comparatively recent years that the true altitudes have become known.

The story of these peaks begins with a daring and historic bit of mountaineering. On May 1, 1827, the Hudson's Bay express, after starting at daybreak and plodding through the snow for seven miles, had stopped at noon just below the summit of Athabaska Pass for the first meal of the day. Though the march through forest and deep snow had been arduous, the vigorous young botanical "passenger" with the express was not played out. The stupendous mountains which at a distance had so awed him were now all around him; one was right at his elbow, seeming to challenge his strength and determination. What followed is best told in Douglas's own words:

After breakfast at one o'clock, being as I conceive on the highest part of the route, I became desirous of ascending one of the peaks, and accordingly I set out alone on snowshoes to that on the left hand or west side, being to all appearance the highest. The labour of ascending the lower part, which is covered with pines, is great beyond description, sinking on many occasions to the middle. Half-way up vegetation ceases entirely, not so much as a vestige of moss or lichen on the stones. Here I found it less laborious as I walked on the hard crust. One-third from the summit it becomes a mountain of pure ice, sealed far over by Nature's hand as a momentous work of Nature's God. The height from its base may be about 5500 feet: timber, 2750 feet; a few mosses and lichen, 500 more; 1000 feet of perpetual snow; the remainder, towards the top 1250, as I have said, glacier with a thin covering of snow on it. The ascent took me five hours; descending only one and a quarter. Places where the descent was gradual, I tied my shoes together, making them carry me in turn as a sledge. Sometimes I came down at one spell 500 to 700 feet in the space of one minute and a half. I remained twenty minutes, my thermometer standing at 18°; night closing fast in on me, and no means of fire, I was reluctantly forced to descend. The sensation I felt is beyond what I can give utterance to. Nothing, as far as the eye could perceive, but mountains such as I was on, and many higher, some rugged beyond any description, striking the mind with horror blended with a sense of the wondrous works of the Almighty. The aerial tints of the snow, the heavenly azure of the solid glaciers, the rainbow-like hues of their thin broken fragments, the huge mossy icicles hanging from the perpendicular rocks with the snow sliding from the steep southern rocks with amazing velocity, producing a crash and grumbling like the shock of an earthquake, the echo of which resounding in the valley for several minutes. On the rocks of the wood were *Menziesia caerulea; Andromeda hypnoides; Lycopodium alpinum; L.* sp. unknown to me; dead stems of *Gentiana nivalis; Epilobium* sp., small; *Salix herbacea; Empetrum nigrum,* fruit in a good state of preservation underneath the snow; *Juncus triglumis; J. biglumis,* with a few *Musci, Jungermanniae* and lichens.

Wednesday, 2nd.—My ankles and knees pained me so much from exertion that my sleep was short and interrupted. Rose at 3 A.M. and had fire kindled; thermometer 20°.[1]

Such is the account of the first ascent of any of the northerly Rocky Mountains. For a long period it was the only ascent. Not for thirty-one years was another ascent made.[2]

On his way home to England, or after his arrival, Douglas thought of giving names to the mountain he had climbed and to another close by. In a revision of his journal he says of the

[1] Douglas, *Journal,* pp. 258-259.

[2] Other peaks were surmounted by Dr. (afterwards Sir) James Hector in 1858-59. To the south, Pike's Peak in Colorado had been scaled in 1820.

former: "Its height does not seem to be less than 16,000 or 17,000 feet above the level of the sea," adding that vegetation ceases at 4,800 feet (above the Pass), beyond which is "1200 feet of eternal ice." He omits the statement in the first journal that he saw many higher mountains, and concludes: "This peak, the highest yet known in the Northern Continent of America, I felt sincere pleasure in naming 'Mount Brown,' in honour of R. Brown, Esq., the illustrious Botanist, a man no less distinguished by the amiable qualities of his mind than by his scientific attainments. A little to the southward is one nearly of the same height rising into a sharper point; this I named 'Mount Hooker,' in honour of my early patron, the Professor of Botany in the University of Glasgow. This mountain, however, I was not able to climb." [3]

Brown is famous as the first Keeper of the Botanical Department of the British Museum, having been appointed shortly before Douglas's return to England. "The greatest botanist of his day . . . , it is doubtful if any greater intellect than Brown's has ever been devoted to the service of Botanical Science." [4] He was presented with a set of Douglas's plant specimens soon after Douglas reached England.[5]

In a further revision of his journal the height of Mount Brown is said to exceed 6,000 feet from its base and 17,000 feet above sea level.[6] Thus Douglas made the height of Athabaska Pass about 11,000 feet.

It is one thing to climb and name a mountain, and another to make it known to the world. The latter Douglas intended doing by publishing the revised journal (probably with a map of his travels), but the plan came to nought. However, another method presented itself.

Hooker was then engaged upon his great work, *Flora Boreali-Americana,* incorporating the recent discoveries of Douglas,

[3] Douglas, "Sketch," *OHQ,* VI, 210-211.
[4] F. W. Oliver, ed., *Makers of British Botany,* p. 3.
[5] Letter, Oct. 29, 1827, Sabine to Brown, British Museum.
[6] Douglas, *Journal,* p. 72.

Drummond, and Richardson. A map showing their travels was to accompany it. Of necessity, Douglas assisted in preparing this map.[7] Here was his opportunity. In addition to marking his routes, he saw to it that prominently displayed in the right spot were the words and figures: "Mt. Brown 16000, Mt. Hooker 15700." That was all that was necessary. Although the professor had misgivings about the figures (writing to Richardson, who also collaborated: "I only wish Douglas had left out his

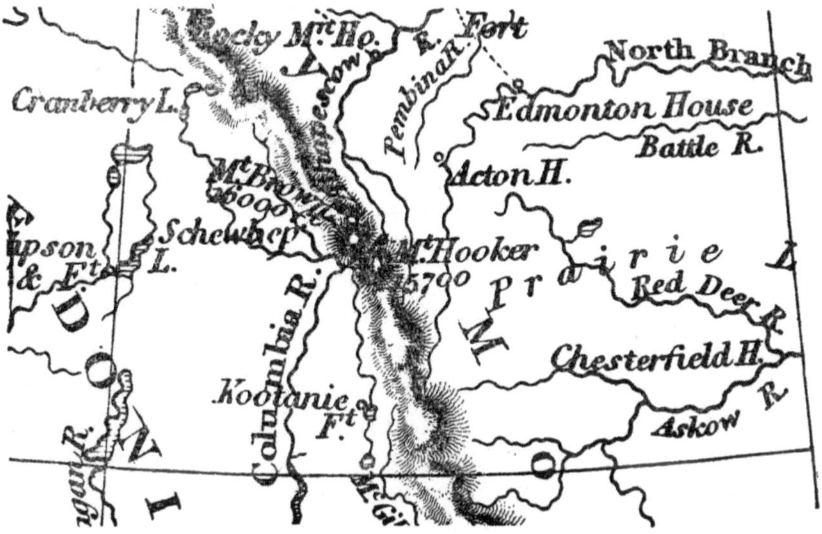

MOUNT BROWN AND MOUNT HOOKER *as shown on the map published in W. J. Hooker's* Flora Boreali Americana *(London, 1840)*

Mount Brown and Mount Hooker, which he has surely most egregiously overrated as to height"),[8] he did not interfere. Accordingly, the map was completed and was published with the first part of the book in 1829.

In this manner, with the sanction of no less an authority than the British Admiralty (where the map was prepared), were constituted and proclaimed to the world—and copied by the

[7] Letter, Aug. 6, 1829, Douglas to W. J. Hooker, Kew Gardens.
[8] Letter, Jan. 22, 1829, Kew Gardens.

map-makers of the period and of generations after—the two mountain monarchs which reigned supreme throughout the whole of North America for more than seventy years!

In due time news of Douglas's mountain-naming and mapping reached the fur traders at Fort Vancouver, probably from his own lips on his return there in 1830. (He had a copy of the map with him.) They do not appear to have been particularly interested. To them mountains had little attraction; instead, they were obstacles to be avoided.

After some years, however, they were obliged to take an interest in the matter. In 1836 Washington Irving's *Astoria* was published, the appendix to which contained statements attributed to the noted explorer David Thompson that the mountains in the vicinity of Athabaska Pass "were nearly as high as the Himalayas," and that "by the joint means of the barometric and trigonometric measurement, he had ascertained the height of one of the peaks to be about twenty-five thousand feet, and there were others of nearly the same height in the vicinity."[9] This information appears to have interested the Royal Geographic Society, for soon afterwards that organization, with the permission of the Hudson's Bay Company, caused carefully graduated thermometers, fitted with apparatus for ascertaining altitudes by the boiling temperature of water, to be distributed among the Company's officers throughout its territories with an earnest request that they use them and report the results.[10]

Accordingly, during an express journey across the Rocky Mountains in 1839, Dr. McLoughlin took thermometrical observations with boiling water at Athabaska Pass and obtained an elevation of 7,324 feet; and he afterwards warned Lieut. Wilkes of the United States Exploring Expedition against ac-

[9] See also *David Thompson's Narrative of his Explorations in Western America, 1784-1812,* J. B. Tyrrell, ed. (Toronto, 1916), pp. 447-448; *Canadian Alpine Journal,* IX (1918), 45-53; Burpee, *On the Old Athabaska Trail,* pp. 68-69.

[10] *Journal of the Royal Geographic Society,* IX (1839), 121-124.

cepting Douglas's figures.[11] Another officer of the Company, Dr. W. F. Tolmie, while eastbound in 1841 estimated the height of the Pass at 7,000 feet.[12] Chief Factor James Douglas, however, was not convinced by his associates, but adhered to Thompson's figures as given in *Astoria*.[13] The information which was received by the Geographic Society from the fur traders led Richardson, who had been instrumental in the matter, to report in 1851 that the peaks of the Rocky Mountains rose from 12,000 to 15,000 feet above the sea, and the passes from 6,000 to 8,000 feet; but Douglas's estimates for mounts Brown and Hooker were not directly called in question.[14]

In those early days opinions regarding the heights of mountains were, of course, largely guesswork and often wrong. Mount Hood (11,245 feet), for instance, close to Fort Vancouver, the headquarters of civilization in the Far West, was the object of miscalculation for many years. In 1843 Rev. Gustavus Hines, the Oregon missionary, declared its height to be 16,000 feet.[15] In 1854 William Barlow, leader of the first ascent party, fixed it at 18,361 feet; while later in the same year another climber made it 19,400. Lieut. Broughton of Vancouver's expedition, who discovered and named the mountain in 1792, generously gave it 25,000 feet.[16]

Meanwhile mounts Brown and Hooker continued in their glory. Extravagant altitudes might be attributed to other mountains, such as Hood, but they were not put on the maps as were those of the Athabaskan peaks. Support was given by the pub-

[11] Charles Wilkes, "Diary of Wilkes in the Northwest," Edmond S. Meany, ed., *Washington Historical Quarterly*, XVI (1925), 290, 300; Charles Wilkes, *Narrative of the United States Exploring Expedition. During the years 1838, 1839, 1840, 1841, 1842* (Philadelphia, 1845) V, 137.

[12] *Transactions, Oregon Pioneer Association, for 1881* (Salem, 1882), pp. 43-44.

[13] Wilkes, "Diary," p. 290.

[14] *Journal*, Geological Society, London, VII (1851), 212; see also John Richardson, *Arctic Searching Expedition* (London, 1851), II, 162-163.

[15] Hines, *Wild Life*, p. 150.

[16] Mazamas, *Mount Hood*, Oregon Out of Doors series (Portland, 1920), pp. 13-18.

lication of Washington Irving's *Astoria* in 1836 and his *Captain Bonneville* in 1843, both of which contained David Thompson's figures and served to confirm the heights of mounts Brown and Hooker, if not their supremacy.[17] Thus we find Father de Smet a few years later declaring: "Upper Athabasca is, unquestionably, the most elevated part of North America. All its mountains are prodigious"; and Roche de Smet (8,330 feet) he credited with 14,000 feet upon the occasion of its being named after him.[18] On the other hand, Dr. Hector of Palliser's railway survey expedition in 1859 was "inclined to think that none of the Rocky Mountains rise above 13,000 or 13,500 feet," but since he "easily recognized Mount Brown and Mount Hooker" in the distance, their supremacy was not challenged.[19] The map accompanying Palliser's report, while showing them prominently, did not show their altitudes.

With the exception of Douglas's climb of Mount Brown and some climbs by Dr. Hector in 1858-59, the Canadian Rockies remained unconquered until the latter part of the nineteenth century. Explorers like Palliser and railway surveyors like Moberly were interested in finding mountain passes rather than mountain peaks. The Canadian Rockies were too inaccessible for mountaineers; the Alps could be reached much more easily. Hence it was not until the Canadian Pacific Railway reached the foothills in 1884 that the Canadian mountains began to receive the attention they deserved.

After some of the peaks convenient to the railway had been climbed, the supposedly stupendous mounts Brown and Hooker became the great attractions. After two unsuccessful attempts to reach Athabaska Pass (long since abandoned as a trade route) Arthur Philemon Coleman, professor of geology at To-

[17] *Astoria* (Philadelphia, 1836), Appendix; *Adventures of Captain Bonneville* (1843), chap. xxv.

[18] P. J. de Smet, *Oregon Missions*, pp. 197-198.

[19] Palliser, *Journals*, pp. 129, 149.

ronto University, and his party succeeded in 1893. They estimated the height of the Pass at 5710 feet.[20] Mount Brown was climbed and its height estimated at 9365 feet; but they were in doubt as to which of the two adjacent mountains was Mount Hooker, and didn't much care, for their enthusiasm had left them when they found the two mighty peaks had shrunk 7000 feet in height.[21] "Mount Brown and Mount Hooker were frauds and we were disgusted at having been humbugged by them," says Coleman. They returned home wondering how Douglas could have made "so monumental a blunder. . . . That two commonplace mountains, lower by two thousand or three thousand feet than some of their neighbours to the south-east, should masquerade for generations as the highest points in North America seems absurd." [22]

Other mountaineers could not believe that Coleman had found the right mountains; the discrepancy between his figures and those of Douglas and Thompson was incredible; possibly there was another Athabaska Pass; the mountains which had been "attractively mysterious to at least two generations" must be found.[23] So the search was carried on for several years.[24] The Pass had long ago ceased to be used by fur traders, railway surveyors, or anyone else, and was very hard to reach. Finally, in 1920 the Interprovincial (Alberta-British Columbia) Boundary Commission confirmed Coleman's location of Mount Brown and fixed its height at 9,156 feet and Athabaska Pass at 5,751 feet (now shown on the official maps as 5,724 feet). A mountain a few miles easterly across the Pass was determined as Mount Hooker with an elevation of 10,782 feet. These determinations

[20] A. P. Coleman, *The Canadian Rockies, New and Old Trails* (London, 1911), pp. 170-218.

[21] *Ibid.*, pp. 202, 206-207.

[22] *Ibid.*, pp. 203, 208.

[23] Stutfield, Hugh E. M. and J. Norman Collie, *Climbs and Explorations in the Canadian Rockies* (London, New York, and Bombay, 1903), pp. 70-71.

[24] J. M. Thorington, *The Glittering Mountains of Canada* (Philadelphia, 1925), pp. 165-167.

were adopted by the Geographic Board of Canada, and thus the matter was settled—officially.[25]

But it was not allowed to rest. A few years later (1925) Dr. J. Monroe Thorington of Philadelphia, after visiting the Athabaska Pass region and making the first ascent of Mount Hooker, and after visiting England to examine Douglas's journals, published a book describing his climbs in the Rockies and dealing with the history of Athabaska Pass and mounts Brown and Hooker.[26] He disagreed with the location of Mount Hooker by the Boundary Commission and contended that the mountain so named by Douglas was McGillivray's Rock (8780 feet), just above the Pass on the opposite side from Mount Brown. He also doubted Douglas's sincerity and suggested that in the revision of his journal his ambition may have led him to add fiction to actuality.[27]

This is a serious charge, and it requires some consideration.

First of all, it must be repeated that Douglas's estimate of the altitude of the mountains above their base at Athabaska Pass is not so much in error as his estimate for the Pass itself. Mount Brown (9,156 feet) he put at 16,000 feet, and Mount Hooker (10,782 feet) at 15,700 feet, while Athabaska Pass (5,751 feet) he put at 11,000 feet. Thus his overestimate of Mount Brown is nearly all accounted for by his overestimate of the Pass; similarly, his overestimate of Mount Hooker is more than accounted for. If he had been fictionizing or guessing, would he not have done it with regard to the altitudes above the base instead of the altitude of the base itself? There would have been much less likelihood of being caught in the former than in the latter. The trail through the Pass was then traveled regularly, whereas the mountains might not be climbed for many years—and in fact were not, as we have seen (Mount Brown in 1893, Mount Hooker not until 1924).

[25] *Canadian Alpine Journal,* XII (1921-22), 163-169; *Report of the Interprovincial Boundary Survey Commission,* part II, pp. 88-89, 94-97.

[26] Thorington, *The Glittering Mountains.*

[27] *Ibid.,* pp. 155-159, 294-302.

The question, then, narrows down to why Douglas so overestimated the altitude of Athabaska Pass. Thompson's figures were not published until 1836. Upon what information did Douglas base his estimate?

For the answer we must go back to Fort Vancouver. During the winter of 1826-27 one of his messmates was Lieutenant Aemilius Simpson, R.N. He had just come overland from Hudson Bay, bringing a set of instruments with which he took the latitudes and longitudes of places on the way. The magnificent scenery and tremendous heights of the Rocky Mountains impressed him, and, assisted by George Barnston, he spent a forenoon in measuring the altitude of one of the first that he approached, Roche Miette (7599 feet). This he estimated to be 3755 feet above the level of Jasper Lake (3287 feet), but, since he considered that the altitude of the lake above the sea "must be very great," he concluded (and recorded in his journal) that the height of the mountain "at a very moderate computation . . . cannot be less than 13000 feet above the level of the sea, and . . . is not much above half the height of some of those more in the interior of the Mass, as I found the more we penetrated into the Mountains the higher they became." [28] Thus he put the altitude of Jasper Lake at upward of 9200 feet—an error of 6000 feet (greater than Douglas's overestimate of Athabaska Pass).

No doubt these estimates were made known to Douglas in the course of conversation during the long winter evenings. The overland route was to be his way home in the spring, and his inquiring mind would lead him to obtain all the information that the Lieutenant could give him regarding the journey. The figures had the stamp of authority, coming from an officer of His Majesty's Royal Navy and backed by instrument calculations. They probably stirred the young naturalist and gave rise to the desire which months later led him up Mount Brown in a mistaken belief as to its height.

[28] Aemilius Simpson's 1826 Journal, MS., Hudson's Bay Co. Archives, London.

After leaving the mountains, Douglas met Thomas Drummond, botanist of Franklin's expedition, who traveled with him from Fort Carlton to Norway House and from Hudson Bay to England. No doubt they discussed their respective adventures and experiences, among which Douglas's mountaineering exploit would have been included. Drummond had accompanied Lieutenant Simpson through the mountains the previous year and was familiar with his opinion in regard to their elevation. "He thinks," wrote Drummond afterwards, "that the altitude of the Rocky Mountains may be stated at about 16,000 feet above the sea." [29] This estimate he apparently accepted without question. His attitude must have strengthened Douglas's belief.

If any doubt remained it was probably removed at Norway House, where Douglas met Governor Simpson. He too had an exalted notion of the height of the mountains, having estimated McGillivray's Rock (now McGillivray's Ridge, 8,780 feet) to exceed 13,000 feet when he first crossed through Athabaska Pass in 1824.[30]

But neither Governor Simpson nor Lieutenant Simpson was the first to overestimate the Rocky Mountains. An earlier exaggeration was made by Ross Cox while crossing eastbound in 1817. He put the Pass at 11,000 feet, but the mountains were too much for him: "They are covered with eternal ice and snow, and will probably be forever inaccessible to man." [31] Possibly all these overestimates may be traced back to David Thompson, who about 1820 or 1822 made the statements mentioned in *Astoria,* and may have offered similar estimates prior to leaving the West in 1812. A great explorer and an experienced land surveyor, his opinion could not well be doubted. Arrowsmith, the famous London map publisher, had a very high opinion of Thompson's knowledge of the Hudson's Bay Company's territories;[32] and it is noteworthy that his maps showed mounts

[29] Drummond, "Sketch," p. 190.
[30] Merk, *Fur Trade,* p. 35.
[31] Cox, *Adventures,* II, 208-209.
[32] *Proceedings,* Royal Geographic Society, London, III (1859), 320.

Brown and Hooker prominently, with Douglas's elevations.[33]

The answer, then, to Thorington's charge is that Douglas was neither fictionizing nor guessing when he overestimated the altitudes of Athabaska Pass and mounts Brown and Hooker. He naturally accepted the well-established opinion as to the height of the Pass, and used that as the basis for his estimates of the mountains. What else could he have done? He had neither the qualifications nor the instruments necessary to check Lieutenant Simpson's estimate. A few years later, when possessed of both, his observations elsewhere were excellent for their approach to accuracy.

Douglas has also been criticized for stating that the trees in Athabaska Pass grew at an elevation of 11,000 feet and that trees grew for 2,750 feet higher. It is said that his knowledge of the altitude limits of vegetation should have warned him against accepting such figures.[34] This, of course, is a criticism made in the light of present-day knowledge. To find trees growing at elevations hitherto unheard-of was no more peculiar than other botanical curiosities that he discovered. Nor did scientists question it, although the statement was published as early as 1836. Instead we find the eminent naturalist, Sir John Richardson, writing in 1851 that the "*R. chamaemorus*, cloudberry, . . . is found on the summits of the Rocky Mountains between latitudes 52° and 56°," [35] whereas in fact these peaks—the highest in the Canadian chain—are without vegetation whatever, being perpetually snow-covered.

Two years after the publication of his book, Thorington took the hundredth anniversary of Douglas's ascent of Mount Brown as the occasion for further and harsher criticism, charging that Douglas "knowingly falsified in claiming the peaks as the loftiest on the continent," and that the naming and estimating were

[33] See map of British North America accompanying *The Canadian North-west, Its Early Development and Legislative Records*, E. H. Oliver, ed. (Ottawa, 1914).

[34] *Canadian Alpine Journal*, IX (1918), 50, 52; XVII (1928), 68.

[35] John Richardson, *Arctic Searching Expedition* (London, 1851), II, 292-293.

done for his own glorification. He even doubted whether Douglas ascended to the summit of Mount Brown, and he criticized the Interprovincial Boundary Survey Commission for its location of Mount Hooker.[36] This brought a reply from one of the two commissioners, Arthur O. Wheeler, F.R.G.S., founder and for many years president of the Alpine Club of Canada, who pointed out that Douglas's words, "the highest yet known," probably referred to his own ascent and naming of the mountain; that he (Wheeler) had climbed Mount Brown twice and had no hesitation in saying it could be climbed in five hours, the time stated by Douglas; that the location of Douglas's Mount Hooker is a matter of surmise; and that the Commission's decision may be just as correct as Thorington's opinion in favor of McGillivray Rock.[37] This was followed by a rejoinder from Thorington upholding his criticisms.[38]

Against Thorington's hypercritical attitude there is the broadmindedness and tolerance of Wheeler, expressed in his concluding paragraphs as follows:

> A number of Dr. Thorington's strictures seem to be his own personal views, and he does not appear to make allowance for the midwinter aspects of unknown solitudes for which there were absolutely no data available, or for the effect of these vast and ice-bound regions upon a solitary traveller. Travelling through them in mid-summer, with available data for position and height ready to hand in the form of reliable government topographical maps, is quite a different thing, furnishing as it does every opportunity for criticism at ease.
>
> The writer, for one, prefers to make due allowance for the abnormal conditions under which the explorations by these first of all pioneers were made, all the more so because in the absence of definite knowledge much in the records is a matter of conjecture.

[36] J. M. Thorington, "The Centenary of David Douglas' Ascent of Mount Brown," *Canadian Alpine Journal,* XVI (1926-27), 185. In regard to Mount Hooker, see Burpee, *On the Old Athabaska Trail,* pp. 159-160.

[37] Arthur O. Wheeler, "Mounts Brown and Hooker," *Canadian Alpine Journal,* XVII (1928), 66-68.

[38] *Ibid.,* pp. 69-70.

There are certainly discrepancies that seem inexplicable in the light of latter-day knowledge, but discrepancies as to the magnitude and effect of many natural features and causes are frequently found among the writings of the early explorers, and are not unlikely due to a magnified impression of the dangers and terrors of mountain travel, especially in mid-winter, before such travel became an every day happening to summer tourists.

Douglas, it must be remembered, had no mountaineering equipment: no ice-axe or alpenstock, no colored spectacles, no edge-nailed boots, and no companion.

The charge of self-glorification seems unfounded. The mistakes in altitude were made innocently and upon the best, if not the only, authority, and the figures were accepted by the British Admiralty and the leading map-makers. In the naming no ulterior motive appears. What could have been more fitting and proper than for Douglas to give to the mountain he climbed the name of the most eminent botanist of the day, and to the other close by the name of his teacher and benefactor, just as he gave their names to plants he found? [39] The charge would have some weight if he had named a mountain after himself; and he might easily have done so, for he saw plenty of them. His own name could have been put on the map as well as the others—an honor which was to come fifty years after his death, when a peak northeast of Lake Louise was named after him by Dr. G. M. Dawson, Director of the Geological Survey of Canada (Mount Douglas, 11,017 feet).[40]

[39] *Paeonia Brownii.* A forget-me-not (*Myosotis*) was named after Hooker but not published.

[40] There is another mountain named after him in Oregon (Mount David Douglas, 6253 feet, in the Cascade Range).

12

FAME AND ITS AFTERMATH

The Horticultural Society had sent Douglas to northwest America in the hope that "from this expedition, our garden will become as well filled with the beautiful vegetation of the borders of the Columbia, and of the Rocky Mountains, as it is already with that of the Ohio and Mississippi."[1] This hope was realized in the first lot of plants received from him (April, 1826). Assuming that this comprised the whole collection, Sabine had written him: "All you have done has given perfect satisfaction and my only wish about you is to see you safe again in England."[2]

The accounts of his hardships, and the fates of previous collectors, made the Society's officers gloomy about the probability of his return. "Our collector," wrote President Knight to his daughter, "proposes, when he has sent all he can home by a ship, to march across the continent of America to the country of the United States on this side, and to collect what plants and seeds he can in his journey; but it is but too probable that he will perish in the attempt. Mr. Sabine says, that if he escapes, he will soon perish in some other hardy enterprise or other. It is really lamentable that so fine a fellow should be sacrificed."[3]

A year later, however, the receipt of another grand lot of plants and reassuring news of his travels lessened the fears for his safety. Sabine wrote him: "You will find us on your arrival not only delighted to see you but ready to make every acknowl-

[1] *Transactions,* Horticultural Society of London, VI (1826), v.
[2] Copy of letter, June 10, 1826, Sabine to Douglas, Royal Horticultural Society.
[3] Knight, *Papers,* p. 39.

edgment to you for your exertions and for your pains and labour."[4]

Now the heroic collector for whom there had been such apprehension was safe home; and to make everybody supremely happy he had brought with him far more plants and seeds than he had sent.[5] Indeed, he had set a record by introducing into England more plants than ever before had been introduced by any individual from any country.[6] Moreover they were of peculiar value and interest for being hardy enough to bear the English climate without protection in winter.[7] The botanical world was literally startled by the number and importance of his discoveries.[8]

The Horticultural Society gave him an enthusiastic reception. Sabine reported to the Council "that Mr. Douglas had throughout his mission acted in the most satisfactory manner and that nothing could surpass the zeal and spirit with which he had executed the trust he had undertaken."[9]

He actually swamped them with seeds. There were too many for the garden, so the oversupply was ordered to be distributed amongst nurserymen. The surplus seeds of the two trees which it was proposed to call *Pinus Lambertiana* and *Pinus Douglasii* were apportioned, first, to Fellows who had applied for them; secondly, to the principal nurserymen; and thirdly, "to such Fellows of the Society in small quantities as will be most likely to plant them on their Estates and to give them the fairest chance of succeeding since it is supposed that (especially with *P. Lambertiana*) it will be a matter of difficulty hereafter to

[4] Copy of letter, March 10, 1827, Sabine to Douglas, Royal Horticultural Society.

[5] *Transactions*, Horticultural Society of London, VII (1830), ii.

[6] "Catalogue of Works on Natural History, lately published, with some notice of those considered the most interesting to British Naturalists," *Magazine of Natural History* (Loudon's), London, I (1828), 58.

[7] *Transactions*, Horticultural Society of London, VII (1830), ii.

[8] Maxwell T. Masters, in *Journal of the Royal Horticultural Society*, XIV (1892), 14.

[9] Minutes of Council, Horticultural Society of London, March 1, 1828.

obtain more Seeds."[10] It was further "ordered that Seeds at present and plants hereafter of such of Mr. Douglas' importations as have been or shall be raised in the Garden of the Society be presented to the public gardens of the Kingdom."[11]

Possibly the enthusiasm was greater from the fact that the entire cost of the expedition, including Douglas's remuneration, was under £400.[12] According to George Bentham, "His whole expenses for food, etc., whilst amongst the Indians, three years, amounted to £66 including a wager of £5 he lost to an Indian Chief. He used to pay the Indians by drafts on the Hudson's Bay Co. for little articles such as a few nails a scalpel knife beads etc. which makes the detailed account transmitted by the Co. to the Horticultural Society a very curious document."[13]

One species of shrub alone, it was said, justified the entire expense of the expedition—the beautiful red-flowering currant (*Ribes sanguineum*), which soon embellished the gardens of Europe.[14]

Douglas was very diffident regarding his success. Before leaving America he had written Governor Clinton:

> In Botany my expectations have not been realized, but at the same time, being in possession of several not included in the American Flora, many interesting and but partially known species, with some additional knowledge as to the geographical range of plants, an enquiry of the greatest importance, I have no reason to regret the journey. At all events, my humble exertions will I trust convey and enthuse, and draw attention to the beautifully varied verdure of N. W. America.[15]

For some weeks after his arrival in London he was too ill to do much. A paper on his favorite tree, the sugar pine (*Pinus*

[10] *Ibid.*
[11] *Ibid.*
[12] George Bentham, "Autobiography," MS., Kew Gardens, vol. II, p. 471.
[13] *Ibid.*
[14] John Lindley, "Ribes sanguineum," *Edwards's Botanical Register,* XVI (1830), plate 1349.
[15] Letter, July 9, 1827, Columbia University.

Lambertiana), which he had prepared for the Linnean Society had to be read by Sabine (November 6, 1827).[16]

As soon as he was able to be about, the young explorer was introduced by Sabine to many of the savants of London, who were greatly interested in his travels and discoveries in far-away and little-known northwest America. They were thrilled by his stories of running for his life from savage Indians, of having to devour the skins of animals he had dried to send to England, and of twice being obliged "to eat up his horse. One of these horses was a favorite whom he called his botanical horse because he had taught it in going through the woods to stop when passing under branches of trees which might tear off the bundles of plants which it carried on its back." [17]

Honors soon came to him. The first was membership in the Linnean Society, the great natural history body which included the world's leading scientists. An accepted nominee was usually ushered into a meeting presided over by the Right Honourable Lord Stanley, wearing a three-cocked hat of ample dimensions and seated in a crimson armchair in great state, who rose and, taking him by the hand, admitted him as a Fellow of the Society. This ceremony usually cost the new member £25,[18] but the Council of the Society "ordered that all Fees payable on the Admission of Fellows, and all Fees payable in future, be remitted in the case of Mr. David Douglas, in consequence of the great services he has rendered to Natural History." [19]

Next honor, with another free fellowship, came from the Zoological Society of London, which, although founded only the year before, already had nearly a thousand members, a museum of several thousand animals or parts of animals, and a menagerie of nearly a hundred live animals.[20] Douglas pre-

[16] *Transactions*, Linnean Society, XV (1827), 497-500.

[17] Bentham, "Autobiography," II, 471.

[18] R. Lloyd Praeger, "William Henry Harvey, 1811-1866," in *Makers of British Botany*, F. W. Oliver, ed., p. 207.

[19] Minutes of Council, Linnean Society, June 24, 1828.

[20] *Magazine of Natural History* (Loudon's), I (1828), 78.

sented the Society with male and female specimens of the California condor which he had shot in the vicinity of Fort Vancouver,[21] and contributed an article on this bird to the Society's *Journal*.[22] He also wrote an article on species of deer and mountain sheep hitherto undescribed (*Cervus leucurus* and *Ovis californianus*).[23]

He was greatly pleased with these honors.[24] Nor were they all. At the annual meeting of the Geological Society of London on February 20, 1829, it was reported that among the most important donations to its museum during the previous year were specimens from the Columbia presented by Douglas;[25] and on June 5 he was elected a Fellow of this Society—again without payment of fees.[26]

This succession of decorations and all that went with them almost turned his head. In the realm of Natural History his name was on everyone's lips. According to his old friend Booth, "his company was now courted, and unfortunately for his peace of mind he could not withstand the temptation (so natural to the human heart) of appearing as one of the *Lions* among the learned and scientific men in London. . . . Flattered by their attention, and by the notoriety of his botanical discoveries, which were exhibited at the meetings of the Horticultural Society, or published in the leading periodicals of the day, he seemed for a time as if he had obtained the summit of his ambition." [27] His discoveries were featured prominently in the Horticultural Society's *Transactions* by means of large colored engravings, for admiration by the most learned and noble not only in England, but abroad as well (the

[21] F. S. Hall, "Studies in the History of Ornithology in the State of Washington (1792-1932)," *The Murrelet*, Seattle, XV (1934), 13-14.

[22] *Zoological Journal*, London, IV (1829), 328-330.

[23] *Ibid.*, pp. 330-332.

[24] Letter, June 14, 1828, Douglas to W. J. Hooker, Kew Gardens.

[25] *Proceedings*, Geological Society of London, I (1834), 106-108.

[26] *Ibid.*, pp. 145, 176.

[27] Letter, W. B. Booth to W. J. Hooker, in W. J. Hooker, "Memoir," *OHQ*, VI, 222.

Emperor of Russia and the crowned heads of the Netherlands and of Denmark and Bavaria were Fellows of the Society).[28] For several months his plants were also the chief attraction in *Edwards's Botanical Register.* This publication, edited by John Lindley, the highly efficient assistant secretary of the Horticultural Society, was the leading botanical periodical.[29] Its beautifully colored plates and apt descriptions were, and still are, highly prized by nature lovers.

Naturally these things stirred Douglas's pride and vanity. His ordinary modesty was stifled by the warmth of his welcome. He lived in an atmosphere of commendation such as is accorded to few.

As further compensation for his services, the Horticultural Society agreed that he should have any profit which might accrue from the publication of his journal, and that he should be given "every assistance and facility . . . which the Society could command or bestow in aiding him therein."[30] John Murray, the London publisher, made him a handsome offer for it,[31] and in anticipation of its acceptance issued an advertisement of the forthcoming publication.[32] Sabine and Lindley offered their assistance,[33] but Douglas rejected it and undertook to prepare the journal entirely himself.[34] For some reason the plans for publication were not carried out. This was disappointing not only to him but to scientists generally. Hooker, in particular, was put out, for he had hoped to get botanical information for his own publications. In a letter to Richardson deploring Douglas's procrastination he sums up the situation thus: "He has much in his head but is totally unfit for authorship."[35] Douglas, however, put the blame on Sabine and com-

[28] *Transactions,* Horticultural Society of London, VII (1830).
[29] *Magazine of Natural History* (Loudon's), I (1828), 58.
[30] Minutes of Council, March 1, 1828.
[31] W. B. Booth, in *Gardener's Magazine,* London, XI (1835), 272.
[32] W. B. Booth, Memorandum on David Douglas, MS., Kew Gardens.
[33] W. J. Hooker, "Memoir," *OHQ,* VI, 222.
[34] *Ibid.*
[35] Letter, Sept. 13, 1828, Kew Gardens.

plained most bitterly that he was the means of withholding his material from the public.[36]

In 1836 a memoir containing extensive selections from the journal was published by Hooker,[37] and for many years this constituted practically the only source of information regarding Douglas's western American travels.

Apparently the journal that Douglas had hoped to publish was to have been a revision by him of the more detailed journal which was finally published by the Royal Horticultural Society in 1914.[38] He did succeed in preparing an abstract of his travels to be used as a preface to his account of the American pines,[39] but, like the journal, both abstract and account remained unpublished until 1914.[40]

It was not long before Douglas tired of being lionized. While he was gratified by the honors conferred on him, his restless nature got little satisfaction from being patronized by the scientists and nobility of England. "When the novelty of his situation had subsided," says Booth, "he began to perceive that he had been pursuing a shadow instead of a reality." [41] By the latter part of April, 1828, six months after his return, he was wearied with it all, and was "dying to go out again upon some expedition." [42] While North America had its dangers, at the same time it had its freshness and freedom; it had its hardships but it also had its opportunities. For the naturalist it was a storehouse of treasure.

To show his distaste for London society and his longing to

[36] Letter, Oct. 20, 1828, W. J. Hooker to Dr. John Richardson, Kew Gardens.

[37] W. J. Hooker, "A Brief Memoir of the Life of Mr. David Douglas, with Extracts from His Letters," *Companion to the Botanical Magazine*, London, II (1836), 79-182 (reprinted in *Hawaiian Spectator*, Honolulu, II, 1839, and in *Oregon Historical Quarterly*, V and VI, 1904-1905).

[38] Douglas, *Journal*. Further in regard to Douglas's journals, see Bibliography.

[39] Letter, June 22, 1829, Douglas to Joseph Sabine, Royal Horticultural Society.

[40] Douglas, Journal, pp. 50-76, 338-348.

[41] Letter, W. B. Booth to W. J. Hooker, in W. J. Hooker, "Memoir," *OHQ*, VI, 222.

[42] Bentham, "Autobiography," II, 471.

return to the wilderness he resorted to untidiness and surliness, becoming "quite a sauvage in his appearance and manners." [43] He wanted to go farther south than he had been before and to explore California, but it was reported that the Indians there were so ferocious that the Society did not like to send him.[44] Moreover, no further expedition was contemplated, the Society's "extensive correspondence with every accessible country now rendering such a means of procuring Horticultural novelties less important than it has been heretofore." [45]

This was a serious disappointment to Douglas's sensitive nature, and he became peevish and irritable. His discontent was increased by finding that after all his exertions and achievements he was not as well paid as the Society's porter. In addition, he discovered that the skins of some birds and other natural history specimens which he had preserved with great care and sent home in 1825 as a grand present for Sabine had not even been unpacked, and were either destroyed by moths or totally unfit for description, and thus were irreparably lost to science.[46] Well-nourished, indeed, was his vexation.

Much of his time was spent in the Society's garden, which was luxuriant with the first results of his recent expedition.[47] There he could forget his troubles. Those plants, sprung from seeds garnered by his own hands with such toil and sacrifice, now bravely pushing up their stems amid the smoke and grime-laden atmosphere of their new habitat, gave him real satisfaction. They brought back many recollections of his days in America. He watched over them with almost parental care. In a very few years they became the most extensive and valuable collection in the ornamental department of the garden.[48]

From time to time he selected plants to send to his revered

[43] *Ibid.*
[44] *Ibid.*
[45] *Transactions,* Horticultural Society of London, VI (1826), v.
[46] W. B. Booth, Memorandum.
[47] Bentham, "Autobiography," II, 439.
[48] *Transactions,* Horticultural Society of London, VII (1830), ii.

Dr. Hooker at Glasgow. They were featured in the *Botanical Magazine* and the *Flora Boreali-Americana,* both of which were the professor's publications. When it appeared that nothing was coming of Douglas's efforts to publish his journal, Hooker visited London (April, 1829) and obtained the Horticultural Society's permission to publish descriptions of all of his plants and to use his manuscripts and specimens.[49] In fact, the material for the *Flora,* probably Hooker's greatest work and for many years the leading work on Canadian botany, consisted chiefly of his pupil's collections and those of Richardson and Drummond.[50] The map accompanying the book[51] was originally drawn showing Douglas's route tinted in yellow, but Douglas persuaded the Admiralty to change it to green, "for yellow is a most sickly hue for a *culler of weeds.*" [52]

The Glasgow professor had a wonderful knack of obtaining new plants from all parts of the world. While Douglas was in London, Captain F. W. Beechey arrived home bringing rare plants from the Pacific, including some from California. They went on to Hooker.[53] A few days later Nathaniel Wallich, superintendent of the Calcutta Botanical Garden, arrived with "an immense store of treasure," according to Douglas, who thought the "prodigious ferns" were "worth coming 100 miles to see." [54] This too all went to Glasgow.

Douglas was delighted that these rarities were going to his teacher and patron, and in the course of his frequent letters to him teasingly expressed his delight: "*Everything goes to Glasgow. Dr. Hooker gets all,* is in everyone's mouth." [55] "Where

[49] Letter, April 29, 1829, W. J. Hooker to Dr. John Richardson, Kew Gardens.
[50] W. J. Hooker, "Memoir," *OHQ,* VI, 218-219.
[51] See Chapter 11; also letter, Aug. 6, 1829, Douglas to W. J. Hooker, Kew Gardens.
[52] Letter, Sept. 14, 1829, Douglas to W. J. Hooker, in W. J. Hooker, "Memoir," *OHQ,* VI, 226.
[53] Letters, July 1, Nov. 1 & 18, 1828, Douglas to W. J. Hooker, Kew Gardens.
[54] Letters, July 18, Nov. 18, 1828, Douglas to W. J. Hooker, Kew Gardens.
[55] Letter, Nov. 18, 1828, Douglas to W. J. Hooker, Kew Gardens.

can I go to find new plants nowadays, you have them from all corners of the earth, not an island nor a creek but you have some portion of its verdure." [56]

Douglas regarded the doctor as more than a mere tutor and friend, holding him in the esteem which a son has for a revered and beloved parent.[57] Most of his letters to him are signed: "Ever your obliged servant"; some have the ending: "Your obliged and grateful servant," or "Your obliged & attached servant." One concludes: "I feel myself indebted to you, not for much, but for all that I possess." [58]

Just as Douglas's botanical collections constituted some of the chief material for Hooker's *Flora,* so his zoological specimens contributed largely to the great book of Dr. John Richardson, *Fauna Boreali-Americana.* Greater assistance would have been given the latter had not many of the birds gone to decay owing to the length of the voyage and other causes beyond control.[59]

Another book in course of preparation to which he contributed was James Wilson's *Illustrations of Zoology.* It illustrated by large colored plates (life-size whenever possible) animals remarkable for their beauty, scarcity, or peculiarity.[60] Among them were male and female grouse shot by Douglas in the vicinity of Kettle Falls (Richardson's Grouse; *Tetrao Richardsonii, Dendragapus obscurus Richardsoni*), named after Dr. Richardson.[61] These birds and the Franklin grouse shot by Ermatinger in Athabaska Pass[62] were presented to Wilson, who deposited them in the Edinburgh University Museum.[63] They

[56] Letter, Dec. 1, 1828, Douglas to W. J. Hooker, Kew Gardens.
[57] Barnston, "Abridged Sketch," p. 121.
[58] Letters, Douglas to W. J. Hooker, Kew Gardens.
[59] Richardson, *Fauna Boreali-Americana,* II, xiii. For list of birds collected or observed by Douglas in the Columbia River region see Hall, "Studies."
[60] Published first in parts at indefinite periods, and in book form in 1831.
[61] Plates xxx, xxxi.
[62] Douglas, *Journal,* pp. 71, 258.
[63] John H. Stenhouse, "Birds of Historic Interest in the Royal Scottish Museum," *Scottish Naturalist,* Edinburgh, May-June 1930, p. 83; Hall, "Studies," pp. 11-13.

were moved in 1865 to the Royal Scottish Museum at Edinburgh, where the Franklin grouse still is.[64]

One more book must be mentioned. The tree which Douglas encountered so often in northwest America and knew as *Pinus taxifolia* had been illustrated and described in Lambert's great work on the genus *Pinus*[65] from twig specimens brought home by Menzies in 1795. Lambert called it *Pinus taxifolia*, or Nootka Fir. He was unable to give a description of the cones because Menzies's specimens had none.[66] Douglas now supplied him with specimens which had cones. This enabled him to complete his description and plates for a later edition of the book, in which at Sabine's suggestion the name *Pinus Douglasii* was adopted,[67] whence we have the name "Douglas fir." Lambert's book also described the *Pinus Lambertiana* and other Douglas discoveries.

Douglas had many interesting hours with Menzies. No doubt he was proud of having obtained the seeds of trees and flowers of which Menzies, owing to the conditions in which he worked, could only make herbarium specimens.[68] Noteworthy among them were the Douglas fir, the arbutus (or madrona) named after Menzies, the California poppy, and the salal. Regarding the sugar pine Menzies told Douglas that in California in 1793 he had seen the seeds served as dessert by the Spanish priests.[69] A whole day would be spent with Menzies in his garden. In the evening of one such day Dr. John Scouler turned up, much to Douglas's delight: "Such a meeting! The three North West Americans under the same roof. He is to dine with us tonight at the Club as Mr. Sabine's guest. Lord Stanley is to preside." [70]

[64] Letter, May 15, 1936, A. C. Stephen, Keeper, to the author.

[65] Aylmer Bourke Lambert, *A Description of the Genus Pinus* (London, 1803). One of the most sumptuous botanical works.

[66] *Ibid.*, p. 51, tab. 33; also 2nd ed., 1828, I, p. 58, tab. 36.

[67] *Ibid.*, 3rd ed., 1832, I, p. 82, tab. 47-48. The official name for the Douglas fir is now *Pseudotsuga taxifolia.*

[68] C. F. Newcombe, ed., *Menzies' Journal*, p. xix.

[69] Douglas, in *Transactions*, Linnean Society, XV (1827), 497-500.

[70] Letter, June 17, 1828, Douglas to W. J. Hooker, Kew Gardens.

Much of his leisure was spent at the home of William Atkinson, the well-known architect, at Grove End, St. John's Wood. He had supervised the erection of Scone Palace during Douglas's boyhood and, being also a botanist, had come to know the lad. Douglas was always welcome at the Atkinson home, where he was more like one of the family than simply an acquaintance.[71]

During Douglas's sojourn in London the Oregon boundary question became acute, and he was asked by the Colonial Office for his opinion on the matter. It was quite a compliment for a twenty-nine-year-old plant collector, but there was nobody else in Great Britain who had explored the territory as extensively as he had. He replied with a memorandum briefly describing the Columbia River and the adjacent territory from the sea to Boat Encampment, mentioning the fertile soil, valuable timber, and mild climate. He pointed out the great importance of the river owing to the immense territory which it drained, and concluded: "There is not any natural Boundary which could give a plea to the American Government to claim this fine country up to 49°. Neither have they priority of discovery either on the Coast or in the Interior. The Boundary Line ought to extend (from my observation on the spot) from the '*Lake of the Woods*' keeping the same parallel to the *Rocky Mountains*, from thence south on the Eastern base, to the pass of *Lewis and Clark 46° N. Lat. and then cross over the dividing side of the Continent to the sources of 'Solomon River' which stream those travellers descended untill they came on the Columbia in 46°00 37″ N. Latt. 119° W. Long. and from that point to the sea on the Columbia leaving the river open to both powers.*" [72]

In this he supported a suggestion which had been put forward by the Hudson's Bay Company.[73] Probably he had discussed

[71] Booth, Memorandum.

[72] Letter, Nov. 27, 1828, Douglas to R. W. Hay, and accompanying memorandum, Colonial Office Papers, class 6, volume 6, folios 807-813, Public Record Office, London. Douglas's mention of Solomon River means Salmon River.

[73] Letter, Dec. 9, 1825, J. H. Pelly to George Canning, *Oregon Historical Quarterly*, XX (1919), 31-33.

the matter with Deputy Governor Garry, and the request for his opinion may have been prompted by Garry, knowing what the answer would be.

In the autumn of 1828 he visited Hooker at Glasgow, taking with him garden specimens of his "American things," which he discussed at length with the professor, giving him useful information for his forthcoming book.[74] He was amazed at the wonderful growth of the Glasgow Royal Botanic Garden, which under Hooker's unbounded energy and enthusiasm had been developed in a few years to become the largest (in number of species) in Great Britain, and probably in Europe.[75] Likewise Hooker's herbarium and library were becoming the best privately owned ones in Europe, by reason of his extensive correspondence and judicious purchases and the consignments from former pupils in various parts of the world.

As a result Glasgow was a Mecca for botanists. From far and near they flocked to view these displays of botanical wealth and interest.[76] Owing to his earlier associations, Douglas was profoundly affected, and he was fortified in his desire for greater fields to conquer.

[74] Letters, Aug. 28 and Nov. 1, 1828, Douglas to W. J. Hooker, Kew Gardens.

[75] *Magazine of Natural History* (Loudon's), I (1828), 399-401; Bower, F. O., "Sir William Hooker," pp. 129-130.

[76] J. D. Hooker, "Sketch," pp. xxxii-xxxiii, xxxvi-xxxvii.

13

SECOND GREAT ADVENTURE

As TIME WENT ON Douglas became more restless. He was surfeited with the monotonous praise and the fashionable society, and, like many another who has come under the spell of the wilderness, he was irked by the restraints and congestion of the metropolis and disgusted with its smells, smoke, and grime. In November he was ill for twelve days with a severe cold and sore throat. "The London climate kills me," he wrote Hooker.[1] A year had passed since his return from America, and still nothing was in prospect. He pined with disappointment.

As hope of a further expedition dwindled he became irritable and cantankerous. Even his friends did not escape. He assumed to sit in judgment on them and expressed his conclusions unsparingly. Lindley, who in addition to his other duties had just been appointed professor of botany at London University, and who has been called "the dominant personality in Botany of the early and mid-Victorian era,"[2] had been complimented by Douglas in the naming of *Oenothera Lindleyii.*[3] It was not long, however, before he came under Douglas's criticism. At the dinner commemorating the 200th centenary of John Ray, Douglas thought all the speeches were splendid with the sole exception of Lindley's (a maiden effort and impromptu): "The beginning was bad, the end was bad and the middle worthy of the beginning and the end, not one sentence worth repeating, and the manner of delivery shockingly ill."[4]

Criticism grew into distrust and enmity. Douglas attended a

[1] Letter, Nov. 18, 1828, Douglas to W. J. Hooker, Kew Gardens.
[2] F. W. Oliver, ed., *Makers of British Botany*, p. 3.
[3] *Magazine of Natural History* (Loudon's), I (1828), 165. Now called *Oenothera amoena.*
[4] Letter, Dec. 1, 1828, Douglas to W. J. Hooker, Kew Gardens.

course of lectures by Lindley, some of which were not well attended—"only thirty and I am willing to believe that fifteen are hired to attend," he wrote Hooker.[5] Lindley was anxious to share with Hooker the herbaria of the Arctic travelers. "Certainly for the present I should not give him any," Douglas advised. "I am not certain but I shall manage to keep him out of my collections . . . He shall not have pleasure picking of them as he had before of my former collection and then to toss me overboard when he obtained it all. I trust you will not think me spiteful but really after the usage I have received if you knew all I am confident you would in one moment say I was right, and would advise me to be more cautious for the future." [6]

Nuttall, whose high ideals had impressed him at Philadelphia, he now found, in London, selfish and conceited. "Nuttall is a poor fellow," he told Hooker, " . . . He did not come from his pinnacle of greatness to visit the Linneans though his fees of admission have not yet been paid, much less his annual subscription. . . . Next to an untruth there is nothing in the world I detest more than the narrowmindedness we too frequently see even in men who ought to be above such thoughts. A man of science who labours not for *self* but from an honest endeavour to add to the stock of knowledge must of all things feel an inexpressible delight in beholding his fellowman engaged in the same laudable undertaking. I am not surprised that Mr. Nuttall should think the *Arkansas and Missouri his own,* for I have heard from some who had reason to know him that he does not practice the liberality he professes." [7]

To Nicholas Vigors, secretary of the Zoological Society, he

[5] Letter, June 7, 1829, Douglas to W. J. Hooker, Kew Gardens.

[6] Letter, Aug. 6, 1829, Douglas to W. J. Hooker, Kew Gardens.

[7] Letter, Oct. 24, 1832, Douglas to W. J. Hooker, Kew Gardens. Although not written until then, it probably expressed the conclusions Douglas arrived at in 1828. Moreover, it was written soon after his resignation from the Horticultural Society's service, when he was in a similar frame of mind. Apparently he feared his condemnation was too broad, for he went on to say, somewhat equivocally: "So far as I know Mr. N. personally he appears the reverse and we must not credit all we hear."

sent a curt and petulant note protesting against being styled "Esq." in the Society's list of members and being called "Mr.," and concluding with this fling: "As it is doubtful if I even can afford a residence I have further to request that you give me no address." [8]

Apparently the Horticultural Society had continued him in its employ without any plan for the future. To dismiss him would look shabby, in view of his achievements, and would provoke a storm of criticism. Yet his presence was frequently annoying, and what to do with him must have been a problem.

It was a critical period in his life and it reveals his weakness. Briefly, the circumstances were too much for him. The hob-nobbing with the noble and learned was in itself a strain for one of humble birth and limited educational advantages, especially in those pre-Victorian days of strong class distinctions. He might have weathered that, however, had the outlook for the future been bright. As it was, he had nothing at all in prospect. He might be called a man of science, but he could not expect a position as scientist. The usual reward for one of his vocation was a berth as curator of some public or private botanic garden; but nothing of the sort was in sight for him, and what he really wanted—another expedition—was frowned on by the Society. Accordingly, when he recovered from the adulation that had been showered on him and realized that he was merely a successful plant collector with no hope of advancement or renewed engagement, his restless nature rebelled and led him to make himself more or less of a nuisance. The self-will of his school-days was again showing itself. He lacked the stability and balance for meeting his disappointments with equanimity. The self-reliance that he had developed during his western American travels, while sufficient for that sort of environment, was unsuitable for meeting difficult situations in refined society. The fine art of complete self-control had not been learned, either in the old land or the new.

[8] Copy of letter, Dec. 29, 1828, Douglas to Vigors, Royal Horticultural Society.

Hooker himself, his best friend, became annoyed with his churlish behavior. The kindly mannered professor thus expressed himself:

"Qualified, as Mr. Douglas undoubtedly was, for a traveller, and happy as he unquestionably found himself in surveying the wonders of Nature in its grandest scale, in conciliating the friendship (a faculty he eminently possessed) of the untutored Indians, and in collecting the productions of the new countries he explored; it was quite otherwise with him during his stay in his native land. It was, no doubt, gratifying to be welcomed by his former acquaintances, after so perilous yet so successful a journey, and to be flattered and caressed by new ones; and this was perhaps the amount of his pleasures, which were succeeded by many, and, to his sensitive mind, grievous disappointments. . . . His temper became more sensitive than ever, and himself restless and dissatisfied; so that his best friends could not but wish, as he himself did, that he were again occupied in the honourable task of exploring North-west America." [9]

Hooker had a good opportunity for sizing up the situation during a visit to London in the spring of 1829;[10] and his remarks have greater weight by having been written several years afterwards—subsequent even to Douglas's death.

Hooker's visit seems to have had a steadying effect. "I have been honestly working very hard ever since you left London," wrote Douglas soon afterwards, "have got my Herbarium arranged and placed in the Council room . . . very neatly. The Journal is still far behind but I am doing my best to get finished." [11]

The visit probably also did much more than that for Douglas. Hooker had a high opinion of his former pupil and believed he should be sent out again. He wished he and Drummond could go to California together.[12] At all events, by July the Horticultural Society had changed its attitude and decided to send

[9] W. J. Hooker, "Memoir," *OHQ*, VI, 221-222.
[10] Letter, June 7, 1829, Douglas to W. J. Hooker, Kew Gardens.
[11] *Ibid.*
[12] Letter, Oct. 20, 1828, W. J. Hooker to Dr. John Richardson, Kew Gardens.

Douglas again to western North America, this time with the assistance of the Colonial Office and the Hudson's Bay Company.[13] His salary was to be £120 a year.[14] His field was to embrace the whole territory west of the Rocky Mountains from California northward as far as he could safely get,[15] but his principal object was to discover the botanical treasures of the interior of California and to make them known in Europe as he had those of the Columbia.[16] He was to sail on the 15th of September, on the Hudson's Bay Company's ship *Eagle*, Captain Grave.[17]

With this news his spirits rose. He plunged into preparations for the expedition, and began to fit himself for carrying on a new activity in addition to those of his previous travels.

Sabine had a brother, Captain (afterwards General, Sir) Edward Sabine, who was secretary of the Royal Society, the greatest of all scientific organizations. To him Douglas one day expressed regret that owing to his limited education he was unable to render service to the geographical and physical sciences in keeping with the excellent opportunities afforded by his travels. He felt particularly his inability to determine geographical positions, since the longitudes, latitudes, and directions of the rivers and mountains which might serve as natural boundaries in the territory of the Columbia (which he was about to visit again) would be valuable to the nation in connection with the unsettled boundary. Appreciating the importance of the matter,

[13] W. J. Hooker, "Memoir," *OHQ*, VI, 222; also *Transactions*, Horticultural Society of London, VII (1830), ii, iii. Richardson also may have had a hand in the matter, for Douglas was to send him zoological specimens (letter, Nov. 2, 1831, Richardson to W. J. Hooker, Kew Gardens).

[14] Minutes of Council, Horticultural Society of London, July 1829.

[15] *American Journal of Science and Arts*, XVI (1829), 145-156; Edward Sabine, "Observations taken on the Western Coast of North America by the late Mr. David Douglas, with a Report on his Papers, 1837," MS., Royal Society, London.

[16] Letter, Aug. 6, 1829, Douglas to W. J. Hooker, in W. J. Hooker, "Memoir," *OHQ*, VI, 223.

[17] Letters, Aug. 6 and Sept. 14, 1829, Douglas to W. J. Hooker, in W. J. Hooker, "Memoir," *OHQ*, VI, 223-225.

and knowing something of Douglas's physical and mental vigor,[18] Captain Sabine told him that if his determination was as strong as his desire he could acquire the requisite knowledge in the three months which remained before his departure, and that if he wished to attempt it he would cheerfully undertake to instruct him.

Douglas accepted the kind offer and enthusiastically entered upon a course of arduous and intense study. At that time his knowledge of arithmetic did not extend beyond the first four rules, and he was by no means expert with those. Since they might be mastered during his outward voyage, however, the Captain began instructing him at once in the use of the instruments necessary in fixing geographical positions and the methods of computing the results, including the minimum of plane and spherical trigonometry and logarithms essential for his practical purposes. He got along famously. The energy and determination with which for eighteen hours a day he concentrated upon overcoming his deficiencies, and his capacity for picking up knowledge in various subjects under extreme pressure of time, yet keeping each subject as distinct in his mind as if it were the only one studied, excited his instructor's admiration.[19]

He had little time for peevishness now. "I have had only one or two *very slight outbreakings* as Mr. Sabine calls my fits since I saw you," he wrote Hooker in August.[20]

Four weeks were spent with Captain Sabine at Greenwich Observatory.[21] He learned the uses of the sextant, chronometer, barometer, thermometer, hygrometer, and compass, and also of apparatus for determining the dip, intensity, and variation of the magnetic needle; for it was intended that he should make

[18] Douglas had assisted Captain Sabine during August, 1828, in the latter's efforts to determine the dip of the magnetic needle by taking observations in Regent's Park (*Gentleman's Magazine*, London, February, 1829, vol. XCIX, part 1, p. 159).

[19] Edward Sabine, "Observations."

[20] Letter, Aug. 6, 1829, Kew Gardens.

[21] *Ibid.*

magnetic observations—Sabine's favorite study—as well as geographical.[22]

As the Horticultural Society could not see its way to furnishing instruments for purposes other than its own, a grant of £80 was obtained from the Colonial Office through Under-Secretary of State Hay toward the cost of furnishing Douglas with sets of the above-mentioned instruments and apparatus and the requisite books.[23] In Sabine's opinion this amount was entirely inadequate, and he proceeded to select instruments costing the enormous sum of £231-8-6.[24] No wonder the responsibility for their proper use made Douglas look on the expedition as "not the journey of a commonplace tourist."[25]

The Colonial Office was also to pay the principal part of his traveling expenses and compensate him for the charts he made; and he was assured that any service of public benefit he might perform would be duly considered by the government on his return.[26] Moreover, although he was to travel out by a Hudson's Bay Company ship, letters of recommendation were sent by the Colonial Office to such of His Majesty's ships as might call on the coast of western North America, directing them to give him "such facilities as they may have it in their power to grant him."[27]

Besides the sources of assistance already mentioned there was the Zoological Society, whose contribution included a double-barreled shotgun costing £18, which Douglas took "in

[22] Edward Sabine, "Observations"; W. J. Hooker, "Memoir," *OHQ*, VI, 222-223; letter, Aug. 6, 1829, Douglas to W. J. Hooker, *ibid.*, *OHQ*, VI, 223; letter, Oct. 27, 1829, Douglas to W. J. Hooker, Kew Gardens; Edward Sabine, "Report on the Variations of the Magnetic Intensity observed at different Points of the Earth's Surface," *British Association for Advancement of Science Report for 1837*, London, VI, 27; *Transactions, 1830-1837*, Royal Society, III, 471-472; *American Journal of Science and Arts*, XVI (1829), 145-156.

[23] Edward Sabine, "Observations."

[24] British Colonial Office, 42/265, fol. 353, Public Record Office, London.

[25] Letter, Aug. 6, 1829, Douglas to W. J. Hooker, in W. J. Hooker, "Memoir," *OHQ*, VI, 223.

[26] *Ibid.*; Edward Sabine, "Observations."

[27] Letters, Sept. 22 and 30, 1829, Hay to Barrow, Adm. 1/4243, Public Record Office, London.

preference to three bad ones." [28] On this expedition he was to be well provided with instruments and arms, and he was going under the best of auspices: the British government, the Hudson's Bay Company, the Horticultural Society, and the Zoological Society, and no doubt with the best wishes of the Royal Society, through Captain Sabine.

In his personal equipment he was not so well treated. Although his list of wants was made out with all the care and economy which the nature and extent of the journey would permit, the Horticultural Society's allowance fell short of the required amount by nearly half, and for a time it looked as if Douglas would be obliged to pay the difference himself. He even asked for £40 of his salary in advance for that purpose.[29] It was made up, however, by contributions from private sources.[30]

A more serious difficulty, and one which could not be overcome, was his failing eyesight. An attack of ophthalmia on the Columbia in 1826, followed by snowblindness (probably during his summer mountain climbs) had so injured his eyes that he could not read small type. Being unable to find in London a Bible to take with him having "a good bold legible type," he asked Hooker to obtain one in Glasgow.[31]

After a short illness brought on by hectic rushing about to complete travel arrangements, purchase equipment, and get letters of introduction,[32] and after a farewell visit to his mother at Scone,[33] he sailed from England on October 31, 1829.[34] He

[28] Letter, Oct. 14, 1829, Douglas to Joseph Sabine, Royal Horticultural Society.
[29] Letter, Sept. 9, 1829, Douglas to Joseph Sabine, Royal Horticultural Society.
[30] Edward Sabine, "Observations."
[31] Letter, Aug. 6, 1829, Douglas to W. J. Hooker, in W. J. Hooker, "Memoir," *OHQ*, VI, 223.
[32] Letter, Sep. 9, 1829, Douglas to Joseph Sabine, Royal Horticultural Society.
[33] *Cottage Gardener*, VI (1851), 263.
[34] Minutes of Council, Horticultural Society of London. The Society's *Transactions*, VII (1830), iii, says he embarked on Oct. 26. If so, he did not sail until later, for he wrote Hooker from London on Oct. 27 (W. J. Hooker, "Memoir," *OHQ*, VI, 226).

was very happy. The prospects were pleasing and exciting, and he felt himself "a very lucky fellow" in having so many people interested in assisting him.[35]

During the long voyage round Cape Horn he was busy with his studies and instruments. On his previous voyage to northwest America he and Scouler had used the ship as a floating dissecting room. Now he used it as a floating observatory. "I never suffered an opportunity to pass," he wrote Captain Sabine, "without endeavouring to perfect myself in the use of some instrument. During the whole passage my days were only moments."[36] Daily observations of the latitude and longitude were taken. They were entered and calculated on lithographed forms bound in volumes brought for the purpose of so recording his travels. He purposed to become so proficient that his land observations in America, after being recomputed on his return to England, would furnish the fixed points for a map of his journey, which would be completed from field sketches of each day's route.[37]

The instruments were of the best quality and worked satisfactorily. Arago, the famous French astronomer, had furnished approved asbestos thread for suspending the magnetic bars. Captain Sabine was familiar with such things and chose them well.[38] The chronometers gave with great exactness the various points at which the ship stopped, and were tried out on land at the Sandwich Islands. When a meridian observation of the

[35] Letter, Sept. 14, 1829, Douglas to W. J. Hooker, in W. J. Hooker, "Memoir," *OHQ*, VI, 224.

[36] Letter, Oct., 1830, in Edward Sabine's "Observations."

[37] Edward Sabine, "Observations."

[38] So well that the cost of the instruments led to a long dispute over who was to pay for them. It continued for years—until after Douglas's death—and it remains a question who paid the bill. It is certain that eight years later, when one of the merchants tried to collect from the government, he was met with a denial of liability and the suggestion that he look to Douglas's representatives for payment (Letter, April 24, 1837, John Booth to Colonial Secretary; copy letter, April 29, 1837, Lord Glenelg to Booth, Colonial Office, 42/277, fol. 221-223, 225, Public Record Office, London).

sun could not be obtained Douglas used double altitudes and frequent meridian altitudes of the stars.[39] The apparatus for determining the variation, force, and dip of the magnetic needle he suspended from the main stay of the ship, where the motion and the influence of metal were least; but the smallness of the vessel greatly hampered him and "almost precluded hope of correctness."

When the ship called at Honolulu he had better success with barometric measurements of the altitudes of several of the peaks on the island of Oahu, and he climbed Mauna Parrii, the seat of the great Akua or God of Fire. He was the guest of Madame Boki, the governess of the island, who entertained him "splendidly." As an especial present for Hooker he obtained a set of all the books that had been published in the Islands, elegantly bound in tortoise shell. He regretted that the ship did not call at Hawaii so he could have ascended the great volcanoes, Mauna Loa and Mauna Kea. The field for botanical collecting intrigued him and he hoped for another and a longer visit.[40]

On June 3, 1830, he arrived again on the Columbia.[41] Although he had made some enemies during his previous stay,[42] the fur traders were glad to see him again and "pleased to find that his stature as a disciple of science had greatly increased," as Barnston says.[43] A new acquaintance, who was struck by his thinning hair and aging appearance, described him as "a fair florid partially bald-headed Scotsman of medium stature gentlemanly address about 48 years of age." [44]

Several changes had taken place during his absence. Fort Vancouver had been moved to a position more convenient to

[39] Letter, Oct., 1830, Douglas to Edward Sabine, in "Observations"; letter, Oct. 11, 1830, Douglas to W. J. Hooker, in W. J. Hooker, "Memoir," *OHQ*, VI, 290-292.

[40] *Ibid.*

[41] Letter, Oct. 11, 1830, Douglas to W. J. Hooker, in W. J. Hooker, "Memoir," *OHQ*, VI, 291.

[42] Letter, Jan. 28, 1830, Richardson to W. J. Hooker, Kew Gardens.

[43] Barnston, "Abridged Sketch," p. 267.

[44] Roberts, "Recollections," p. 8. He was not yet 31.

the river.[45] The garden, under the supervision of Billy Bruce (another Scot), was greatly improved,[46] and the farm extended.[47]

He also found new faces, including several young boys who had been sent from the Greenwich Naval School to be trained for the Hudson's Bay Company's coast service. One of them, George B. Roberts, assisted Douglas with his instruments. The work was mutually beneficial: the boy found it helped to "furbish up my school acquirements." [48] He rose to be treasurer and probate judge of Wahkiakum County in Washington.[49]

After many nights' vigil Douglas determined the position of Fort Vancouver as 45° 37′ Lat. and 122° 36′ Long.[50] The modern determination is 45° 40′ Lat. and 122° 30′ Long. His inaccuracies were probably due to discrepancies in the magnetic observations which he could not account for. Similar inexplicable differences were found later by Captain Belcher and Lieutenant Wilkes.[51]

He then accompanied the interior brigade up the river as far as Fort Walla Walla, taking observations en route and botanizing with his usual eagerness. On this journey and on his later travels in northwest America he took with him William Johnson, an employee of the Company, who acted as his personal servant or "man." [52] Johnson was an old man-o'-war's man with a roving disposition and was well suited for the work. He settled down afterwards and is remembered as the first permanent resident

[45] Bancroft, *Northwest Coast*, II, 438-439.

[46] Hubert Howe Bancroft, *History of Oregon* (San Francisco, 1886-1888), I, 9.

[47] Barnston, "Abridged Sketch," p. 270.

[48] Roberts, "Recollections," p. 8.

[49] Bancroft, *Oregon*, I, 38-39; Hubert Howe Bancroft, *History of Washington, Idaho and Montana* (San Francisco, 1890), p. 42.

[50] Edward Sabine, "Report on the Variations of Magnetic Intensity," pp. 29, 30; Sabine, "Observations." The longitude was arrived at by Sabine after recalculation of Douglas's figures. Further, see Chapter 15.

[51] Wilkes, *Narrative*, V, 123.

[52] Hudson's Bay Co. Transfer Book B 239 CC 1, 1825-1831, Columbia Supplies, fol. 423-425, Hudson's Bay Co. Archives.

within the site of Oregon's metropolis, the city of Portland.[53]

Barnston, who was with the party (having been appointed in charge of Fort Walla Walla), was surprised at Douglas's enthusiasm and "the quickness of sight he displayed in the discovery of any small object or plant on the ground over which we passed. When in the boats, as they proceeded along, he would frequently spring up abruptly in an excited manner, and with extended arms keep his finger pointed at a particular spot on the beach or the shelving and precipitous rocks, where some new or desirable plant had attracted his notice. This was the signal to put on shore, and we would then be amused with the agility of his leap to the land, and the scramble like that of a cat upon the rocks to the object he wished to obtain, happy if he achieved this without slipping, and falling into the deep water alongside the boat." [54]

Douglas made Walla Walla one of his principal stations for astronomical and magnetic observations. He found it a very favorable place for this work owing to the clear sky, low humidity, equal diurnal variations of temperature, and the regularity of barometric and magnetic figurings.[55]

With a guide and horses provided by Barnston he made another journey to the Blue Mountains, his purpose being to study plant distribution at various altitudes.[56] "I again found my *Poenia,*" he wrote Hooker,[57] referring to the *Paeonia Brownii* he had found there four years earlier. He regretted that the wild disposition of the Shoshone Indians prevented his penetrating farther south in their territory.[58]

Upon his return he and Barnston amused themselves with a novel and successful method of catching lizards:

[53] Roberts, "Recollections," p. 11; Henry E. Reed, "William Johnson," *Oregon Historical Quarterly*, XXXIV (1933), 314-323.

[54] Barnston, "Abridged Sketch," p. 268.

[55] *Ibid.*

[56] *Ibid.*

[57] Letter, Oct. 11, 1830, Douglas to W. J. Hooker, in W. J. Hooker, "Memoir," *OHQ*, VI, 291.

[58] Barnston, "Abridged Sketch," pp. 268-269.

Indian boys were employed to beat about for the game armed with single horse-hair lassos, tied to the end of a wand. It was laughable to see the little urchins, naked as they were born, scouring about, and when they discovered a hole, throwing themselves flat on the heated sand, and extending the small noose over the entry to the reptile's apartment. Where their victim shewed his head, they would then quickly suspend him with one jerk, and bring him like a culprit to our sides: a slight reward would put them in ecstasies, and they would again scamper off for renewed captures. The most common species obtained was an *Agama,* the *Tapaya Douglasii,* and a very beautiful long tailed little lizard of a light pavonine iridescent hue. It measured about six inches or more, was particularly agile and appeared to great advantage, as it flitted rapidly before the sun from knoll to knoll. It was probably a *Cnemidophorus.*[59]

His thoughts turning toward California, Douglas left Walla Walla on July 23 and returned downstream.[60] On a Tuesday evening he reached the Cascades, "the land of my little vain Indian Chumtalia. It was *Hyass*[61] Sunday with him. The whole earth from the east and from the west, from the most distant parts, had congregated to make mirth with him, and to *Patchatch*[62] on the never-to-be-forgotten occasion of the perforating the septum of his young daughter's nose, and piercing her ears. It would have been very ungallant to the young lady, as well as ungracious towards the father, to have pressed him or any of his band, to go with me to the mountain at such a time. I therefore deprived myself of that pleasure and proceeded to Fort Vancouver, where I arrived on Wednesday at kail time. Chumtalia in six days more, comes for me in his large canoe." [63]

In August he ascended the Willamette and Santiam Rivers with A. R. McLeod and took observations at various places. Mount Jefferson, a peak in the neighboring Cascade Range, he estimated at 11,320 feet.[64] This, the first recorded estimate by

[59] Barnston, "Abridged Sketch," p. 269.
[60] *Ibid.*
[61] *Hyass,* "great." [Barnston]
[62] *Patchatch,* supposed to mean "feast" [Barnston. Now "potlatch."]
[63] Barnston, "Abridged Sketch," p. 270.
[64] *Ibid.,* pp. 271-272.

anyone, and probably Douglas's first scientific attempt at altitudes, was considerably out. The height is 10,495 feet. His figures for other mountains in the Northwest have not come to light. Probably they were lost when he was wrecked in the Fraser River in 1833.[65]

As trinkets for the Indians he took four jew's-harps, three pounds of round assorted beads, three bundles of barleycorn beads, four dozen plated coat buttons, and a whole gross of finger rings, the lot having been purchased at the Company's Fort Vancouver store for nineteen shillings and four pence.[66]

During the return journey he had the misfortune to lose all the zoological specimens he had collected, "a dreadful loss, as it contained good things. It is curious; on the 17th November 1826, *I lost everything I had at the same place,* when returning from my southern journey! A *kelpie* or elf is the charm of that stream,[67] so unfortunate to me. Since that time I have made my intended trip to the cascades with my friend Chumtalia. I accomplished all I wished. Poor Chumtalia is since dead. He was blown up by his powder horn which was on his person, and falling on his side, his knife entered about the fifth rib so that he died. He is now laid with his fathers." [68] Douglas was all the more disappointed over the loss of these specimens because he had intended sending them to Dr. John Richardson for publication.[69]

Upon returning to Fort Vancouver he had a touch of the terrible intermittent fever which had broken out among the natives. It depopulated the country to an alarming extent, he informed Hooker: "Villages, which had afforded from one to two hundred effective warriors are totally gone; not a soul remains. The houses are empty and flocks of famished dogs are howling

[65] See Chapter 16.
[66] Hudson's Bay Co. Transfer Book B 239 CC 1, 1825-1831, Columbia Supplies, fol. 420-423, Hudson's Bay Co. Archives.
[67] River Sandiam [Barnston].
[68] Barnston, "Abridged Sketch," p. 272.
[69] Letter, Nov. 2, 1831, Richardson to W. J. Hooker, Kew Gardens.

about, while the dead bodies lie strewn in every direction on the sands of the river. I am one of the very few persons among the Hudson Bay Company's people that have stood it, and sometimes I think even I have got a *shake,* and can hardly consider myself out of danger, as the weather is yet very hot." [70]

Douglas was "ten days in that state, between hope and fear, but never was laid down." [71] Twenty-four of the Company's men died;[72] and for a time business on the Columbia was at a standstill.[73] In October he journeyed down the river to Fort George to pack up his collections for shipment home. Of plants, over one hundred new species had been obtained, including many seaweeds. Three chests of seeds were packed. His dried plants were kept back for another vessel to avoid risking the whole collection in one shipment.[74]

Of the genus *Pinus* he sent a bundle of six beautiful species. One of them, *Pinus grandis* (now *Abies grandis*), a grand ornamental tree and the most rapid grower of all the silver firs, was a new discovery. *Pinus nobilis* (*Abies nobilis*), however, Douglas considered "by far the finest. I spent three weeks in a forest composed of this tree, and day by day could not cease to admire it." [75] Seeds which accompanied the specimens of this tree arrived in good condition and were grown successfully. The young plants were distributed among the Fellows of the Horticultural Society and no sooner began to show the tree's splendor than it leaped into universal favor. Extravagant prices

[70] Letter, Oct. 11, 1830, in W. J. Hooker, "Memoir," *OHQ,* VI, 292.

[71] Barnston, "Abridged Sketch," p. 272, quoting from a letter from Douglas. Elsewhere (p. 270) Barnston says that "all at the establishment were sufferers by the unwelcome visitation, and Douglas was ill like others, but being something of a leech, had an early recovery, and recruited perfectly by following up his wonted healthful perambulations."

[72] *Washington Historical Quarterly,* I (1907), 261.

[73] *Ibid.,* II, 40. Most of the writers on early Northwest doings dwell on the ravages of this malaria epidemic. Further, see *Oregon Historical Quarterly,* XXIX (1928), 144-161.

[74] Letter, Oct. 11, 1830, Douglas to W. J. Hooker, in W. J. Hooker, "Memoir," *OHQ,* VI, 291.

[75] *Ibid.*

were paid for it, fifteen or twenty guineas being not unusual for a fine young plant.[76]

Despite his loss at the Santiam, in zoology Douglas had obtained several new species of quadrupeds, birds, reptiles, and insects, which were valuable additions to Richardson's *Fauna.*

Winter was now approaching; and as his movements were hampered by the fever epidemic and the hostile attitude of the natives,[77] he made plans to leave for California. He had hoped to go by land in order to link up his Umpqua journey in search of the sugar pine, but his experiences there with the natives and the fate of Jedidiah S. Smith's party in 1828 [78] made him hesitate.[79] "You may judge of my situation," he wrote Hooker, "when I say to you that my rifle is in my hand day and night; it lies by my side under my blanket when I sleep, and my faithful little Scotch terrier, the companion of all my journies, takes his place at my feet. To be obliged thus to accoutre myself is truly terrible. However, I fail not to do my best, and if unsuccessful in my operations can make my mind easy with the reflection that I used my utmost endeavours." [80]

Deciding that the safer and better way to California was by water, he left Fort Vancouver by the Company's brig *Dryad* late in November. At Fort George and Cape Disappointment he took geographical observations.[81] He arrived at Monterey on December 22, 1830.[82]

Among his purchases just before leaving Fort Vancouver, be-

[76] Edward Ravenscroft, ed., *The Pinetum Brittanicum* (London, 1884), I, 4.

[77] *Washington Historical Quarterly,* I (1907), 258; Merk, *Fur Trade,* p. 304; *Oregon Historical Quarterly,* XIII (1912), 366-369; letter, Oct. 11, 1830, Douglas to W. J. Hooker, in W. J. Hooker, "Memoir," *OHQ,* VI, 291-292.

[78] See Bancroft, *Northwest Coast,* II, 449-452; Bancroft, *California,* III, 152-160; and Sullivan, *Travels.*

[79] Letter, Oct. 11, 1830, Douglas to W. J. Hooker, in W. J. Hooker, "Memoir," *OHQ,* VI, 292.

[80] *Ibid.*

[81] Edward Sabine, "Observations."

[82] Letter, Nov. 23, 1831, Douglas to W. J. Hooker, in W. J. Hooker, "Memoir," *OHQ,* VI, 293; Bancroft, *California,* III, 146.

sides shoes, articles of clothing, cloth, cotton, lead, shot, cod line, candles, etc., were the following:[83]

		£	s	d
10	lbs Hyson tea	3		
60	lbs loaf sugar	1	15	
9	gals Madeira wine	7	1	
1	Fustian Coattee		1	1
1	pair bedford trousers		14	3
9	galls brandy	2		6
2	large black silk Handkerchiefs		17	6
2	Bernagore silk Handkerchiefs		2	6
1	large moose deer skin		5	8
3	prime chev. skins		6	9
2	pairs Nankeen trousers		11	6
1	Jews harp			2
150	spanish dollars	46	8	1

[83] Hudson's Bay Co. Transfer Book B 239 CC 1, 1825-1831, Columbia Supplies, fol. 420-423, Hudson's Bay Co. Archives.

14

CALIFORNIA

DOUGLAS landed in California at a time of political disturbance. The governor—the first under the new Mexican republic—had resigned over difficulties with the national government, and his successor had not yet arrived. Manuel Victoria did not assume office until more than a month after Douglas came. He was a soldier, stern and strict; and several interviews, spread over many weeks, were necessary before he was convinced of the innocence of Douglas's intentions. Finally, on April 20, the necessary passport[1] was granted, its duration being limited to six months. Douglas was warned against making sketches or plans of the fortifications; but since these consisted only of "a mud wall of three sides, open in the rear, with a breastwork about three feet in height, with rotten platforms for seven guns, the discharge of which would annihilate their remains of carriages," and be "more dangerous to those using them than to those against whom they might have been used," [2] it has been suggested that the intention of the warning was that he "was not to make fun of their helplessness." [3]

The morale of the Californian authority was in as bad a state

[1] "1831—Abril 20—Mont. Victoria. Carta de Seguridad de David Duglas. Concede carta de Seguridad para permanecer y transitar pr 6 meses al súbdito británico natural de Escocia Don David Duglas." Translation: "1831—April 20—Monterey, Victoria. Passport of David Douglas. A Passport to stay and travel for six months is granted to David Douglas, Esquire, a British subject and a native of Scotland." *Archives of California, Departmental Records,* Bancroft Library, California, MS. A-49, tomo IX-XIV (1831-1845), 97. The misspellings are from the original.

[2] Edward Belcher, *Narrative of a Voyage Round the World . . . during the years 1836-1842* (London, 1843), I, 136.

[3] Roberts, "Recollections," p. 8.

as the fortifications. Altogether, Douglas's impression was that "the Mexican territorial government as applied to California is abominable, and that is the mildest word I can use." [4]

But in the rank and file of the inhabitants he found a pleasant contrast. They were a simple, homely, happy, and for the most part peaceful people.

Monterey, a straggling village with beautiful surroundings, was the political, commercial, and social center. Here he took up residence with William Edward Petty Hartnell, an English merchant who had been naturalized as a Mexican citizen and had married a daughter of a prominent Spanish family. Men of science seldom visited California, and for one to make a stay was most unusual. Accordingly, as an "hombre de educacion" and called "Don David," Douglas became a welcome member of the little community.[5]

A knowledge of Spanish being essential, he soon acquired it. He had waited a long time before benefiting from his study of the language while en route to the United States in 1823. Probably Hartnell assisted, for he was an accomplished linguist, as familiar with Spanish as with his mother tongue.

Although nearly four months passed before he received his passport, Douglas was not idle. Spring—which comes very early in California—already was at hand, and the Monterey peninsula has the reputation of presenting a greater diversity of flora than anywhere else in the world. The great Del Monte forest with its thousands of acres of beautiful woodlands was in its pristine grandeur.

The first flower plucked was the beautiful wild gooseberry (*Ribes speciosum*), which struck him as "a flower not surpassed in beauty by the finest Fuchsia." It was not a new discovery, having been found by Menzies and taken home to England by Collie, naturalist with Beechey's expedition which visited

[4] Barnston, "Abridged Sketch," p. 275.

[5] Information given the author by Maria Antonia Field, San Francisco, a great-granddaughter of Estevan Munras, a Monterey pioneer.

California briefly in 1826. On the same day, however, he did make a real discovery in the charming little herbaceous annual, baby-blue-eyes (*Nemophila insignis*) which he called "a humble, but lovely plant, the harbinger of Californian spring," and which was destined to become one of the most popular flowers among garden lovers the world over. Other discoveries and astronomical observations kept him busy in the vicinity of Monterey during January.

In February he went east and north to visit the missions of San Juan Bautista, Santa Clara de Assise, and Santa Cruz.[6] Just as the Hudson's Bay Company's forts were the only civilized spots in the Columbia region, so the Franciscan missions (with their schools, farms, vineyards, and workshops) were the outposts of enlightenment in California. The way, therefore, to see the country was to visit the missions, traveling from one to the other—a day's journey on foot—by the old El Camino Real (Royal Road or King's Highway). They were the only stopping places for travelers, and welcomed with generous hospitality the few who came, making no charge for food, bed, or horses.[7]

On this journey he was delighted to find another pine. Near the San Juan Bautista Mission he ascended the Gabilan Range to get the latitude and longitude of the summit,[8] and it was probably here that he made the discovery,[9] (actually a rediscovery of a species found on his journey to the Umpqua in

[6] Known dates of Douglas's itineraries during 1831 are as follows:

Monterey, January 29.	Purisima, May 5.
San Juan Bautista, February 14.	Santa Ynez, May 6.
Santa Cruz, February.	Santa Barbara, middle May.
Santa Clara, February 20.	Monterey, late June, July 15.
Monterey, March 2, 27, April 21.	San Rafael, July 27.
Soledad, April 25.	San Francisco Solano, July 29.
San Antonio, April 27.	San Jose, August 5, 9.
San Miguel, May 1.	Monterey, Aug. 25, Nov. 23.
San Luis Obispo, May 3.	

[7] *Century Magazine,* New York, XLI (1891), 163-183, 389-402.

[8] Letter, Nov. 25, 1833, Douglas to Governor Figueroa, Bancroft Library.

[9] Jepson, *Silva,* p. 90.

1826 and lost in crossing the Santiam River). In spite of its small size, he called it "a noble new species," [10] and "one of the most beautiful objects in nature." [11] He gave it the name *Pinus Sabinii* in honor of Joseph Sabine of the Horticultural Society, and from San Juan he sent Sabine specimens of the cone and leaves together with a description of the tree, which was read before the Linnean Society. He told Sabine he hoped it would "exist and florish when we shall cease to be when we shall be gone forever." [12] Seeds were planted at Kew Gardens, but the tree has not done well in the English climate.[13] (*Sabinii* was changed to *Sabiniana,* its present scientific name. The common name is Digger pine or gray pine. The Digger Indians used the large, sweet, oily seeds or nuts as food and chewed the gum.)

North of the Santa Cruz Mission was the great forest of redwood trees (*Sequoia sempervirens*) which is now the California State Redwood Park. This tree had been discovered by Menzies and described in Lambert's *Genus Pinus* under the name *Taxodium sempervirens.*[14] Nothing more was heard of it until Douglas reported that "the great beauty of California vegetation is a species of *Taxodium,* which gives the mountains a most peculiar, I was almost going to say awful, appearance—something which plainly tells that we are not in Europe." He collected specimens and seeds, but for some unknown reason the tree was not introduced into Europe until several years later. His remark has led some persons to think that he referred to the famous big tree of California (*Sequoia washingtoniana,* but that is a mistake, because his travels took him nowhere near its habitat.[15]

In addition to his botanical activities there were his astro-

[10] Letter, Feb. 4, 1831, Douglas to Joseph Sabine, Royal Society.
[11] *Transactions,* Linnean Society, XVI (1833), 747.
[12] Letter, Feb. 4, 1831, Douglas to Joseph Sabine, Royal Society.
[13] Jepson, *Silva,* p. 89.
[14] Hansen, "Pinetum Danicum," pp. 309-310.
[15] *Ibid.*

nomical observations. He carried his instruments with him on this journey—as he did on his other travels in California—and took observations to fix the latitudes and longitudes of the various missions.[16]

Returning to Monterey early in March, he busied himself thereabouts until late in April. Then, having obtained his passport, he journeyed southward, visiting the missions of Nuestra Senora de la Soledad, San Antonio de Padua, San Miguel Arcangel, San Luis Obispo de Tolosa, La Purisima Concepcion, and Santa Ynez Virgin y Martyr; and arrived at Santa Barbara in the middle of May.

Returning by the same route, with occasional side trips to the near-by mountain valleys, he reached Monterey in the latter part of June.

About the middle of July he traveled northward, visiting San Francisco, San Rafael Arcangel, San Francisco Solano, and San Jose de Guadalupe. Probably from San Jose or San Francisco, he went on a botanical visit to the Mount Diablo region, penetrating as far east as the inner south coast range.[17]

He hoped to go north far enough to reach the place whence he had turned back on his journey southward from Fort Vancouver in search of the sugar pine. However, his most northerly observation, 38°45′, indicates that he did not get much beyond the Solano Mission or the Russian settlement at Fort Ross. The intervening gap of some three hundred miles he erroneously estimated at sixty-five miles (possibly he was misled by finding more of the sugar pines), but even that was too much for him. There was then no well-marked trail through northern California, nor were there any facilities for travel. To have attempted the journey with any degree of safety would have

[16] List in Bancroft, *California,* III, 404.

[17] Willis Linn Jepson, *A Flora of California* (Berkeley, 1936), II, 37; W. L. Jepson, "David Douglas in California," *Madrono,* Journal of the California Botanical Society, Berkeley, Cal., II (1933), 97; Alice Eastwood, "Early Botanical Explorers on the Pacific Coast and the Trees They Found There," *California Historical Society Quarterly,* XVIII (1939), 341.

required a much larger party than he had the means of commanding.[18] He returned to Monterey late in August.

As the months passed by he realized that the season for botanizing was short, consisting of only the spring months. "Nothing can exceed the beauty of a California Spring," he wrote, but "such is the rapidity with which spring advances . . . the plants bloom here only for a day. The intense heat set in about June, when every bit of herbage was dried to a cinder." The torrid heat and the dryness of the atmosphere amazed him, and he lamented "the want of such majestic rivers as the Columbia." Further in the same vein he wrote as follows:

The climate though warm is healthy, and were it not for the intense drought of July and August, the soil would be very productive. In no part of the world have I experienced such a dryness in the atmosphere, nor can I call to my memory having read of greater. Even the deserts of Arabia and Egypt, the plains of Sin and Ispahan in Persia, I mean the driest places on the globe, when satisfactory observations have been made, are more humid than California. Often when the thermometer Fahr., stands at 80° or 100°, 30° or 40° of dryness is by no means unfrequent. On some occasions I have sunk the thermometer below *zero,* and after repeated trials, with all the care I am capable of bestowing on such a delicate operation, not the least particle of moisture could be detected. But nature ever kind and varied in her operations, compensates for this extreme dryness of the daytime by copious dews during the night, at all times proportionate to the dryness of the preceding day. Otherwise animals or vegetables could not live; the most would only be existence, and that for but a short period. In 1831, the rain was only .700 of an inch, the 39th part of the mean of the English climate. Notwithstanding these great drawbacks to this beautiful country, it is the land of the vine, the olive, the fig, the banana and in the southern parts, of the sugar cane, and a variety of the usual fruits seen in semi-tropical climates. The vine is cultivated to a large extent, from 10,000 to 100,000 in one vineyard. The wine is excellent, indeed, that word is too small for it; it is very excellent." [19]

[18] Barnston, "Abridged Sketch," p. 273.
[19] Barnston, "Abridged Sketch," p. 274.

The sun's heat and glare aggravated his eye trouble. Writing to Hooker from Monterey in November he said: "I can never read what I write, so do pardon my blunders and if you can fathom what I wish to say I am for once happy." However, his ability with hook and line was not affected, for he added: "Tell Joseph[20] I caught two fine trout yesterday, twenty seven pounds each."[21]

Travel facilities were very poor. It was a case of riding horseback or walking, for the El Camino Real in some places was only a trail. Most of Douglas's journeys were made on foot, and alone.[22] There were no roads or trails through the interior, much of it being desert, and the weather was unbearably hot for many months of the year. The heat was said to be so intense that it suffocated the fleas.[23] Then, too, he found that "the expense is not the least of the drawbacks." He therefore gave up the idea of visiting the Colorado River region, which he had in mind before leaving England, and decided to renew his labors on the Columbia, taking ship direct or via the Sandwich Islands, whichever vessel came first.

Plant collecting being over for the year, he spent the time in arranging and cataloguing what he had. There were about five hundred species, many of them new, and making several new genera.

Some of his discoveries have been erroneously credited to Dr. Thomas Coulter who arrived in California about this time (November, 1831). Coulter had been plant collecting in Mexico for his old master, De Candolle of Geneva, and had extended his field to California under the impression that it was still unoccupied.[24] He was as surprised to meet Douglas as the

[20] Hooker's son (later, Sir) Joseph Dalton Hooker, 1817-1911.

[21] Letter, Nov. 23, 1831, Kew Gardens.

[22] William Heath Davis, *Seventy-five Years in California* (San Francisco, 1929), pp. 206-207.

[23] George Wharton James, *In and Out of the Old Missions of California* (Boston, 2nd ed., 1927), p. 241.

[24] Letter, March 27, 1835, Coulter to W. J. Hooker, Kew Gardens.

latter was to meet him. But friendship, instead of rivalry, ensued. Douglas welcomed Coulter as another devotee of science, "a good fellow, most zealous in all he undertakes and anxious to collect a vast collection";[25] and it was "a *terrible pleasure* . . . thus to meet a really good man and one with whom I can talk of plants." They went fishing and shooting together and became very chummy—so much so that Douglas prevailed upon Coulter to share some of his discoveries with him so he could send them to his old master, Hooker.[26] Coulter was the first botanist to visit the Colorado desert and Gila region (Douglas's original objective), and introduced into Europe a host of plants, including a great variety of cacti. Apparently he had more means for travel than Douglas. Another visiting naturalist whom Douglas met was Ferdinand Deppe, a German, who was also engaged in mercantile pursuits and traveled about as supercargo.

Although in daily expectation of a ship, Douglas carried on his natural history pursuits by making short trips through the territory near Monterey. Quadrupeds and birds received his attention. A collection of the bones of animals, and of a sea otter intact, was made up and sent to Dr. Scouler. Efforts were made to find a nest of the California condor. On obtaining a specimen of the bird at Fort Vancouver, Douglas had been wrongly informed that the eggs were jet black[27] (they are pale blue, spotted brown), and he was curious to find some. He was unsuccessful however, owing probably to the misinformation, although he carried on the search in all his travels in California.[28]

His scientific pursuits did not make him oblivious to the business activities of the territory. He noted that "vast numbers of Sea Otter are caught on the coast of California—in 1831 not less than 5,000 which is great for the persons employed. Beaver

[25] Letter, April 15, 1833, Douglas to W. J. Hooker, Kew Gardens.
[26] *Ibid.*
[27] Douglas, *Journal*, p. 241.
[28] *Hutchings' Illustrated California Magazine*, San Francisco, 111 (1850), 541; see also *The Murrelet*, XV (1934), 13.

are rare and the fur appears poor. They sell it by weight. Land otter in the interior parts of California are common. The trade is principally carried on by Americans who receive in return merchandize. Hides & Tallow is the wealth of California." [29]

During his stay several parties of fur hunters arrived from St. Louis.[30] Douglas found them "generally intelligent and kind-hearted men, much disposed to be useful" to botanists. They reciprocated by regarding him "as half an American, having spent so many years in the New World."

Days, weeks, and months passed, yet no ship came. A new year (1832) arrived, bringing another Californian spring. Undismayed by the delay, Douglas converted it into an opportunity for more botanizing: "I continued to consider California *as still new to me,* and set to work a second time finding new plants, and drying better specimens of those which I formerly possessed." He could not venture far from Monterey, fearing a ship might come and go without him, so restricted his journeys to fifteen or twenty days. "However, as I did my best, I try to feel content."

Now came activity of a different sort.

A Spanish governor once called California "the most peaceful and quiet country in the world." While that may have been true at the time, it was not so under the Mexican republic, and at the time of Douglas's visit conditions were far from peaceful. In January a threatened revolt caused the foreign residents of Monterey to organize as a "Company of Foreigners" to assist in keeping order at the capital. Nearly fifty men of various nationalities were enrolled, among them Douglas and Coulter, with Hartnell elected captain. They did guard duty in and around the town. Strict discipline was kept; spirituous liquors and gambling were forbidden.[31]

Although some of his comrades were runaway sailors and

[29] Letter, Oct. 26, 1832, Douglas to Joseph Sabine, Royal Society.
[30] *Ibid.*
[31] Monterey Company of Foreigners, orderly book, MS, Bancroft Library.

men of that sort, others were men of substance. There were Captain Juan Bautista Roger Cooper, James (Santiago) McKinley, and Nathan Spear, leading merchants; Jean Louis Vignes, pioneer wine-maker; Thomas Doak, first American settler in California; and George Kinlock, a Scotsman whose wife, also Scottish, was the first foreign woman to live in California. A prominent member was Don Estevan Munras, a Spaniard, who used to entertain the whole company with great Sunday dinners at his palatial home, "La Granja," still one of Monterey's landmarks.[32]

Not enamored of his militia service, Douglas does not even mention it; probably this and his other experiences with Mexican affairs are summed up in his reference to the government as "abominable." The monotony of routine, uneventful guard duty would not suit him. He longed to be on the hunt again; the season for botanizing would soon be gone for another year, so he left for the Santa Lucia Mountains south of Monterey.

Here in March he found another new tree. "You will begin to think that I manufacture Pines at my pleasure," he wrote Hooker. This one was the curious and rare Santa Lucia or bristlecone fir, one of the most beautiful of the silver firs, if not of all the firs. It is unique for its dense, Indian-club-shaped crown, sharp spires, drooping branches, and deep, lustrous, yellow-green foliage of long, flat, sharp-pointed leaves, white-lined underneath. It has a very restricted habitat, being confined to the Santa Lucia Mountains. While it perhaps once covered the summits, the range now is from 2200 to 5000 feet.[33] Possibly he had heard of the tree from Padre Pedro Cabot at San Antonio, for the mission fathers there used the resin for incense, and called the tree *Incensio*.[34] Its distinctive features and its rarity have made it sought after by botanical travelers from all parts of the world. To this "most beautiful object" he gave

[32] Information given the author by Maria Antonia Field, San Francisco, a great-granddaughter.

[33] Sudworth, *Forest Trees*, p. 122.

[34] Jepson, *Silva*, pp. 125-126.

the name *Pinus venusta.* It is now known as *Abies venusta* (charming fir). Sometimes its discovery is wrongly credited to Coulter under the name *Abies bracteata.* Neither Douglas nor Coulter sent seeds home, and it was not introduced into England until 1853 (by William Lobb).[35]

Another pine discovered and introduced into Britain by Douglas is the Monterey pine (*Pinus radiata* Don; *Pinus insignis* Douglas), which is distinguished by its brilliant, deep, grass-green foliage, closely and strongly serrulate leaves, and thick oblique cones. It is the fastest-growing of the California pines. A third California tree he introduced to Europe is the Coulter or big-cone pine (*Pinus Coulteri*).

Not until August, 1832, did Douglas sail from Monterey. By then he had been nineteen months in California. Previous collectors had stayed only for short periods usually at the wrong seasons, and had got merely what was conveniently at hand. Douglas had systematically combed his territory, being not only the first botanist to remain for an extended period, but also the first to leave written descriptions of his gleanings.[36] His collection consisted of hundreds of distinct species, besides seaweeds, mosses, and minutiae. Its extent was equaled only by its novelty and beauty. He hoped it would furnish his "good friend Dr. Hooker with material for two more volumes of the Flora."[37]

There were about sixty new plants, including shrubs and trees. Among them were five mariposa lilies: the lovely white fairy lantern (*Calochortus albus*), the yellow mariposa (*C. luteus*), the golden fairy lantern (*C. pulchellus*), the lilac mariposa (*C. splendens*), and the butterfly mariposa (*C. venustus*). Among several new gilia were the odd birdseye gilia (*G. tricolor*) and the dainty little feather gilia (*G. androsacea*). There were new species of evening primrose and of pentstemon, including the attractive scarlet-bugler (*Penstemon centranthi-*

[35] Hanson, "Pinetum Danicum," pp. 459-461; J. M. Murray, "David Douglas," *Forestry,* London, V (1931), 157.

[36] Jepson, "David Douglas," p. 97.

[37] Letter, Oct. 26, 1832, Douglas to Joseph Sabine, Royal Society.

folius). There were the charming poppy-buttercup-like cream-cups (*Platystemon californicus*), the pretty bush poppy (*Dendromecon rigidum*), and the handsome blazing-star (*Mentzelia Lindleyi*); also a profusion of lupines, and several new brodiaea, poppies, and phacelia, including the tansy-leaved, clover-smelling wild heliotrope (*Phacelia tanacetifolia*). All these beautiful flowers and many others were soon to bloom in European gardens; beside the cultivated perennials of the old land were to grow the wild annuals of the new.[38]

What has been called "perhaps the greatest curiosity of all his collections" was the quinine bush (*Garrya elliptica*),[39] which has been of special interest because it represents a new natural order.[40] But in Douglas's estimation the most singular and interesting plant was the gray-ball sage (*Audibertia incana* or *Salvia carnosa*).

He divided his collection into halves, sending one to the Horticultural Society and the other to Hooker. The arranging and packing of two large bundles for his beloved teacher gave him great pleasure. "Besides, my dear Sir," he wrote, "were this duty left to a London house, they might do it I doubt not by accident, well, while I do it heartily. May you reap equal pleasure looking over them that I had gleaning them and it will be a pleasure beyond what I can express to know you had found something that interests you." [41] Before being packed, the specimens of pines were plunged in warm water in order that the leaves should not fall off.

The discoveries were received in Europe with even greater delight than those from the Columbia. "What a glorious collec-

[38] Lists in Douglas, *Journal*, pp. 326-336; *Transactions*, Horticultural Society of London, 2nd ser., II (1842), 376-377, 440-441; letter, Oct. 23, 1832, Douglas to W. J. Hooker, in W. J. Hooker, "Memoir," *OHQ*, VI, 299. For condensed list see Appendix B.

[39] Charles Francis Saunders, *Western Wild Flowers and Their Stories* (New York, 1933), p. 109.

[40] *Ibid.*

[41] Letter, Nov. 23, 1831, *California Historical Society Quarterly*, II (1923), 223-227.

tion has Douglas sent to me from California!" wrote Hooker to Richardson. "I think I scarcely ever in a collection of such an extent saw so much that is new and rare."[42] This was the more remarkable since some of the seeds and roots had deteriorated during the long voyage, having had to cross the equator twice and been subjected to great extremes of heat and cold. (Some were sent from Monterey, some from Honolulu, and some from the Columbia. Other packages sent via Mexico never reached England, owing probably to the disturbed condition of the republic.)[43]

When the Society's part of the collection arrived its officials were kept busy for some time sorting and arranging the plants, naming those which Douglas had not named, renaming some, preparing descriptions for publication, and planting the seeds and roots in the Society's garden. Descriptions were published in the *Transactions* and in other botanical periodicals.[44] Some of the flowers were pictured in large, beautifully colored plates which, after over a hundred years, are still highly prized by garden lovers. Hooker did not describe them in his *Flora*, for that dealt with British America. They were described, however, in Hooker and Arnott's *Botany of Captain Beechey's Voyage* (London, 1841), the Californian supplement to which is the most important work on them. Many were also included in Torrey and Gray's *Flora of North America* and De Candolle's *Prodromus*.[45]

The reproductions from the roots and seeds planted in the Society's garden were widely distributed, going to foreign as well as home correspondents and to other horticultural societies and botanical gardens.[46] One set went to St. Petersburg, to-

[42] Letter, May 15, 1833, Kew Gardens.

[43] *Transactions*, Horticultural Society of London, 2nd ser., I (1835), 404.

[44] *Edwards's Botanical Register*, *Botanical Magazine*, *Paxton's Magazine of Botany*, all of London.

[45] John Thomas Howell, "A Collection of Douglas' Western American Plants," *Leaflets of Western Botany*, San Francisco, II (1937-1939), 95.

[46] *Transactions*, Horticultural Society of London, 2nd ser., II (1842), 376-377, 440-441.

gether with a set of Douglas's Columbia plants. Only recently these were sent from Russia to a California botanist for study and determination.[47]

Among botanists today the importance of Douglas's California collections is said to be "due not to any intrinsic value in the specimens themselves, but rather to the botanical opinions that have been expressed about them, opinions which were expressed in the form of new species and varieties and which constitute most of the ground work of our western botanical taxonomy."[48] "Hundreds of new species, our most familiar plants, were based on the Douglas collections, study of which is not exhausted even at this day."[49] Botanists have found their study difficult because of the scarcity of information telling where he collected his specimens. He numbered them carefully, but his field journal was lost the next year when he was wrecked in the Fraser River—a serious loss not only to him, but also to botanists ever since.[50]

Douglas acknowledged the great help he had received from the missionaries, whom he called "an excellent class of men,"[51] writing to Barnston as follows:

> I lived almost exclusively with the fathers who without an exception, afforded me the most essential assistance, hospitality to excess, with a thousand little courtesies which we feel and cannot express. I had no bickerings about superstition, no attempts at conversion or the like, the usual complaints of travellers, indeed so much to the contrary, that on no occasion was an uncharitable word directed to me. When there I was under no restraint; my time was entirely my own, feast day and fast all the same, the good men of God gave me always a good bed, and plenty to eat and drink of the best of the land. A more upright and highly honorable class of men I never knew. They are well educated; I had no difficulty from the begin-

[47] Howell, "A Collection," pp. 59-62, 74-77, 94-96, 116-119, 139-144, 170-174, 189-192.

[48] Howell, "A Collection," p. 95.

[49] Jepson, "David Douglas," p. 97.

[50] *Ibid.;* Howell, "A Collection," pp. 61, 94.

[51] Letter, Oct, 26, 1832, Douglas to Joseph Sabine, Royal Society.

ning with them, for saving one or two exceptions, they all talk Latin fluently, and though there be a great difference in the pronunciation between one from Auld Reekie, and Madrid, yet it gave us but little trouble. They know and love the sciences too well to think it curious to see one go so far in quest of grass.[52]

Douglas became well acquainted with some of the priests, referring to them as "my old friends the Fathers," and mentioning particularly Rev. Narciso Duran and Rev. Pedro Cabot, two of the best-beloved missionaries. In a letter to Hooker he took occasion to answer some recent criticisms against the padres:

I have received a copy of Capt. Beechey's book. I entertain a great respect for that gentleman, but I think he has been too severe on the Catholic Missionaries in California.[53] Any man who can make himself well understood by them, either in Castilian or Latin, will discover very shortly that they are people who know something more than their mass-book, and who practise many benevolent acts, which are not a little to their credit, and ought to soften the judgment of the stranger, who has probably had more opportunity of seeing men and things than the poor priests of California. Their errors are the errors of their profession, and I thus make bold to say so, having had reason to know that the individuals in question are honourable exceptions to priests in general. I am no friend to Catholicism, still I should desire to maintain my own opinion without hurting the feelings of others.

He also had a kind word for the California women (whose grace, beauty, and modesty were admired by every visitor), declaring after his return to the Columbia: "The ladies are handsome, of a dark olive brunette, with good teeth, and the dark fine eyes, which bespeaks the descendant of Castille, Catalan or Leon. They (sweet creatures) have a greater recommendation than personal attractions. They are very amiable.

[52] Barnston, "Abridged Sketch," pp. 274-275.

[53] See F. W. Beechey, *Narrative of a Voyage to the Pacific and Beering's Strait . . . in the Years 1825, 26, 27, 28* (London, 1831), II, 16-19, 21-22, 31-33.

On this head I must say, *Finis,* otherwise you will be apt to think, if ever I had a kind feeling for man's better half, I left it in (Calida Fornax) California." [54]

Douglas made many acquaintances during his long stay in California. Among those not already mentioned were Alfred Robinson, pioneer American merchant; Rev. Patrick Short, first churchman from the British Isles to minister west of the Rocky Mountains, and co-founder with Hartnell of the first college in California; Dona Concepcion Arguella, the ill-fated Rezanof's faithful fiancée and early California's most pathetic character; and David Spence, head of a meat-packing establishment.

Douglas was received in California as "Don David." He left it with a different title. One day a boy playing on the deck of a vessel in the harbor at Monterey fell into the hold. He was picked up unconscious and with a broken arm. There being no medical man resident in the village, "Dr. Douglas" was sent for to attend him. Measuring up to the emergency—as he had in the case of the buffalo-mangled fur trader on the far northern prairie—the impromptu doctor set the boy's arm so well and treated him so kindly that years afterwards in grateful remembrance he related the incident in his book of reminiscences.[55]

The new title stuck to Douglas for the rest of his life. And it has continued in use since, both in the Hawaiian Islands and in California. A recent instance is the reference in a history of medicine in California to "that distinguished Scotch doctor—Dr. David Douglas," which credits him with having saved the boy's life.[56]

[54] Barnston, "Abridged Sketch," p. 275.
[55] Davis, *Seventy-five Years,* pp. 206-207.
[56] George D. Lyman, "The Scalpel Under Three Flags in California," *California Historical Society Quarterly,* IV (San Francisco, 1925), 154.

15

THIRD VISIT TO THE COLUMBIA

The ship by which Douglas left Monterey was bound for the Sandwich Islands. The Hudson's Bay Company's ship for the Columbia had not called owing to the death of the captain;[1] so, after waiting until August, and seeing no likelihood of a voyage direct, he embarked for the Islands, hoping to get passage thence to the Columbia.[2]

With him he took nineteen large bundles of dried plants, together with seeds, specimens of timber, etc., and from Honolulu (then Fairhaven), where he arrived on September 7, 1832,[3] he shipped these to England in two chests on the *Sarah and Elizabeth,* a South Seaman of London. The captain, "a worthy little man," placed them in his own cabin for safety. As before, there were liberal quantities for Hooker, in some cases "forty, fifty and some few hundred species that you may be enabled to pay a debt with some friend."[4]

This was the last occasion upon which he divided his specimens between the Horticultural Society and Hooker. Thenceforth the professor was to get all. It came about through an upheaval in the Society, which Douglas learned of at Honolulu.

The Society's affairs had gotten into a bad way. Continual extravagance on a large garden, botanical curiosities, lavish fetes, and costly publications, on the one hand, and neglect in

[1] *Transactions,* Horticultural Society of London, 2nd ser., I (1835), 404.
[2] Letter, Oct. 26, 1832, Douglas to Joseph Sabine, Royal Society.
[3] *McLoughlin's Fort Vancouver Letters, First Series, 1825–38,* E. E. Rich, ed. (London, 1941), p. 338.
[4] Letter, Oct. 23, 1832, Douglas to W. J. Hooker, Kew Gardens.

collecting subscriptions and membership fees on the other, had brought financial embarrassment.[5] The delinquents, it was reported, included "one king (George IV) for his subscription to the garden, 500 guineas; one duke, one marquess, thirteen earls or lords, seventeen honourables or sons of lords, and twelve clergymen." [6] A committee of investigation found that liabilities exceeded assets by over £3000. The membership was torn by dissension.

Under these circumstances Sabine resigned as secretary-manager and was succeeded by George Bentham (a nephew of the philosopher). Things were put on a business basis, economies were introduced, finances reëstablished, and harmony restored.[7] Fetes gave way to auctions, and at one of them 102 plants of *Pinus Douglasii* (Douglas fir), 230 of *Pinus ponderosa* (western yellow pine), and other plants, brought more than £300.[8]

Douglas had heard rumbles of the impending storm before he left England, at which time Sabine's methods were already under criticism. Douglas loyally supported his superior, notwithstanding his differences with him. He even acted as a spy on a recalcitrant member whom Sabine was trying to have expelled.[9] Now, when news of the upheaval reached him at Honolulu, he felt that he must continue loyal, although it meant severing his relations with the Society. Accordingly he wrote on September 9, 1832, resigning his position.[10]

The changes in the Society sank it so low in his estimation that, in writing to Hooker, he called it the "Beastly Club," although he says he did not create the name;[11] while Bentham

[5] *Gardener's Magazine,* VI (1830), 114, 234-252, 369-371.

[6] *Ibid.,* pp. 236-237.

[7] B. Daydon Jackson, *George Bentham,* English Men of Science Series (London and New York, 1906), pp. 66-68.

[8] *Gardener's Magazine,* VI (1830), 506.

[9] *Ibid.,* p. 114.

[10] *Transactions,* Horticultural Society of London, 2nd ser., II (1842), 375; Minutes of Council, Horticultural Society of London, Feb. 12, 1833.

[11] Letter, Oct. 24, 1832, Douglas to W. J. Hooker, Kew Gardens.

was referred to as "a nameless gentleman (??) not far from London," who he suspected would take delight in expunging Sabine's name from the pine named after him by Douglas.[12]

Hitherto pretty free as to his movements, henceforth Douglas was to be entirely a free lance. Hooker, however, continued to be his guiding star, and his sole ambition now was to please his old master by obtaining specimens for his herbarium and publications.

He saw great opportunities round about him in the profuse vegetation of the Islands. In ferns alone there must be five hundred species, he told Hooker. "This may be worth your while to set about a Flora of these Islands which, with what I can furnish as respects the geographical positions, altitudes of the mountains, etc., a handsome something can be made, at all events."[13] He looked forward to a return visit with plenty of time to investigate these matters.

He took a lot of astronomical observations, but the records came to grief in a peculiar manner: "The vile Cockroaches ate up all the paper, and as there was a little oil on my shoes, very nearly demolished them too!" There are several species of this insect in the Hawaiian Islands, and their propensities as household pests have become well known.

Douglas managed to obtain a pair of Sandwich Island geese (now *Nesochen sandvicensis*), large, handsome birds which are peculiar to the Islands and have their habitat high up on the mountainsides. He sent them alive to London, where they arrived safely and were presented to the Zoological Society.[14]

His stay was clouded by a violent attack of rheumatic fever, which prevented him from getting away to the hills and caused

[12] Letter, Oct. 23, 1832, Douglas to W. J. Hooker, Kew Gardens.

[13] Letter, Oct. 24, 1832, *supra.*

[14] Frederick Debell Bennett, *Narrative of a Whaling Voyage Round the Globe from the Year 1833 to 1836* (London, 1840), I, 251; *Proceedings*, Zoological Society (London, 1833), p. 65; William Alanson Bryan, *Natural History of Hawaii* (Honolulu, 1915), pp. 323, 335-336.

much fretting. Determined that a return visit must be made—one long enough to be worth while, for the Sandwich Islands appeared even more interesting than California—he called on the British Consul, Richard Charlton, "a delightful man," and made tentative arrangements for his help.

He left Honolulu on the *Lama,* Captain William McNeill, and arrived at the Columbia on October 14,[15] less than two months after leaving California. It was fast travel for those days. The passage from Monterey to Honolulu had been made in nineteen days on an American vessel of forty-six tons, which led Douglas to comment: "What would have been thought, forty years ago, of passing over more than half of the great basin of the Pacific with such a craft? If steamboats and railroads are not in our way, we, poor wanderers, must take what offers, sometimes good and sometimes bad." [16]

His old Columbian friends were interested to hear of his experiences in California, but apparently he imposed on their credulity. The accounts of his exploits led one of his listeners to remark skeptically that "bears, bulls and tigers had cause to rue the day they went there." [17]

During his two-year absence further progress had taken place at Fort Vancouver. A seventy-ton schooner was under construction on the river bank.[18] Two hundred acres were under cultivation.[19] Both farm and garden were flourishing; grapes had been added to the orchard.[20] Round about were herds of well-

[15] *McLoughlin's Letters,* E. E. Rich, ed., p. 100.

[16] Four years later saw the advent of the *Beaver,* famous pioneer steamship, which puffed up and down the Coast and in and out of its waterways for over half a century (W. Kaye Lamb, "The Advent of the 'Beaver,'" *British Columbia Historical Quarterly,* vol. II, 1938, pp. 163-184).

[17] Letter, Feb. 20, 1833, Archibald McDonald to Edward Ermatinger, Dominion Archives, Ottawa.

[18] "The Correspondence and Journals of Captain Nathaniel J. Wyeth, 1831-6," F. G. Young, ed., *Sources of the History of Oregon* (1899), vol. I, parts 3 & 6, p. 176.

[19] *Ibid.,* p. 177.

[20] Letter, Oct. 26, 1832, Douglas to Joseph Sabine, Royal Society.

fed cattle—four hundred of them—cropping the rich herbage; while horses, goats, and other animals helped to complete the picture.[21]

But there were indications of a much greater change to come. American hunters and traders were forcing their way through the Rocky Mountains and reaching farther and farther west, thereby giving serious concern to the Hudson's Bay Company. One of them had arrived at the very gates of Fort Vancouver. Nathaniel J. Wyeth had come overland from Boston on an ambitious project: nothing less than the establishment of trade throughout the Oregon territory in active opposition to the great Company. A cargo of goods had been shipped round Cape Horn to meet him at Fort Vancouver. It was the beginning of a new epoch. Wyeth was the forerunner of the American invasion of the Columbia.[22]

While waiting for his ship (which never came, being wrecked on the way), Wyeth was generously allowed the privileges of the fort by Dr. McLoughlin. "Eating and drinking the good things to be had there and enjoying much the gentlemanly society of the place," he, of course, met Douglas.[23] On hearing from Douglas of the botanical wealth of the country he became so interested that when he left for Boston early in February[24] he took with him several specimens (probably obtained from Douglas or with his aid), adding more en route, and sent the lot, one hundred thirteen in all, to his friend Thomas Nuttall, whom Douglas had met at Philadelphia. The collection was of such a highly interesting character that Nuttall published an account of it.[25] More important, it induced him to visit the

[21] *Washington Historical Quarterly,* II (1908), 168; III (1912), 235; XXIII (1932), 212.

[22] Bancroft, *Northwest Coast,* II, 557-565, 598.

[23] "Wyeth," F. G. Young, ed., pp. 178, 180-181, 202-203.

[24] *Ibid.*, p. 181.

[25] "A Catalogue of a Collection of Plants made chiefly in the Valleys of the Rocky Mountains or Northern Andes, towards the sources of the Columbia River, by Mr. Nathaniel B. Wyeth, and described by T. Nuttall," *Journal of the Academy of Natural Science,* Philadelphia, VII (1834), 5-60.

Columbia himself two years later with Wyeth's second expedition.[26]

Another mark of progress at Fort Vancouver that winter was the establishment of a school for the half-breed children of the Company's men, with John Ball, one of Wyeth's party, as teacher. This was the first school in the Pacific Northwest.[27] One of the pupils was Ranald MacDonald, eight-year-old son of Chief Trader Archibald McDonald and Princess Sunday (a daughter of Chief Concomly). He afterwards had some remarkable adventures in Japan. The lad became very fond of Douglas, and in later life referred to him as a "dearly loved friend and companion . . . for with all his enthusiasm in his pursuit, he was ever the most sociable, kindly and endearing of men." [28]

The winter was an unusually severe one, the Columbia being frozen across for two weeks with ice six inches thick.[29] The cold weather gave Douglas "an excellent opportunity of multiplying my astronomical observations, on the angular distance between the moon's limb and the sun; the planets Venus, Mercury, Saturn and Mars and the fixed stars: not less than eight thousand observations in about six hundred sets, separately computed for the purpose of ascertaining the absolute longitude of Fort Vancouver." He also observed a beautiful eclipse of the moon on the night of January 5, and many of the eclipses of Jupiter's satellites. He devoted all his skill to these matters in order to obtain the positions accurately, since all his chronometric longitudes were reduced to that meridian. From partially completed calculations he gave a provisional longitude of 122° 39′ 10″ W. He then sent a full report to Captain Sabine, who, after having recalculations made at the Admiralty, con-

[26] *Companion to the Botanical Magazine,* I (1835), p. 15.

[27] John Ball, *Autobiography of John Ball,* compiled by his daughters, Kate Ball Powers, Flora Ball Hopkins, and Lucy Ball (Grand Rapids, Mich., 1925), p. 93.

[28] Ranald MacDonald, *Narrative,* pp. 136-137. Ranald and his father differed in the spelling of the prefix to their name.

[29] "Wyeth," F. G. Young, ed., p. 180.

sidered the true longitude to be 122° 36′ 15″,[30] which still differs somewhat from the present determination of 122° 30′.

With the coming of milder weather in February, Douglas was on the move again. This time it was northward by the Cowlitz River[31] to Puget Sound. His object was to determine the positions of the headlands on the coast and the high mountains in the interior, and their altitudes. His man Johnson knew something of the territory, having gone through the Sound to the Fraser with John Work in 1824.[32]

Although it was a most laborious journey, he was favored with fine weather and succeeded in accomplishing his object. Unfortunately the records of his observations have been lost. His activities included the gathering of about two hundred species of mosses, together with some interesting seaweeds from Puget Sound. Some months earlier, feeling that a "blank" existed "in the department of seaweeds," despite Menzies's collections and his own collections from California, he had written to all his friends up and down the Coast requesting them to collect "everything in the shape of a seaweed" and to dry them and put them in bags for convenience of transport, leaving him to revive them upon receipt. A similar request went to the captains of all ships on the Coast.

Passing near the base of Mount St. Helens on February 26, the prodigious size of some white fir (*Abies grandis*) led him to remark: "A forest of these trees is a spectacle too much for one man to see." [33] This tree, too, he introduced to Europe.

At Fort Nisqually, which had just been established, he met Archibald McDonald early in March, and returned with him to Fort Vancouver soon afterward. He arrived there only in time to hurry off on another journey—one much more difficult and dangerous, and accompanied by disaster.

[30] Edward Sabine, "Observations."
[31] Barnston, "Abridged Sketch," p. 273.
[32] *Washington Historical Quarterly*, III (1912), 200.
[33] Letter, April 9, 1833, Douglas to W. J. Hooker, Kew Gardens.

16

ILL-FATED JOURNEY

"J'ai appris avec une vive joie vôtre intention de faire une tournée dans nos environs. Soyez sûr, Monsieur, que jamais visite ne m'a été plus agréable, et que des bras ouverts vous attendent à Sitka. Si vous avez l'intention de retourner en Europe, par la Siberie, je puis vous assurer qu'au mois de Mai de l'année prochaine, vous pourrez commodement aller sur un de nos navires a Okotsk, où, d'après des nouvelles que je viens d'apprendre, on vous à déjà préparé un gracieux accueil."—Baron Wrangel, Governor of the Russian Territories in America, to Douglas.[1]

BEFORE LEAVING ENGLAND, Douglas had conceived the ambitious idea of returning home from America by Siberia, thus traveling around the world. His purpose was to compare the vegetation and the astronomical phenomena of America with those of similar latitudes in Asia. He also wished to inspect the platinum mines which had just been discovered in Siberia.[2]

"What a glorious prospect!" he wrote Hooker, "Thus not only the plants, but a series of observations may be produced, the work of the same individual on both Continents, with the same instruments, under similar circumstances and in corresponding latitudes! I hope I do not indulge my hopes too far. . . . People tell me that Siberia is like a rat-trap, which there is no

[1] Translation:
"I am delighted to learn of your intended journey to our region. Let me assure you, sir, that never has a visit given me more pleasure, and that you will be received with open arms at Sitka. If you intend to return to Europe via Siberia, I can promise that in May of next year you will be able to go comfortably on one of our warships to Okotsk, where, I have just heard, they have already planned a hearty welcome for you."

[2] Thomas Beale, *The Natural History of the Sperm Whale* (London, 1839), p. 362.

difficulty in entering, but from which it is not so easy to find egress. I mean at least to put this saying to the test. And I hope that those who know me know also that trifles will not stop me."

He submitted the proposal to the Russian authorities through their minister in London, where it fell on sympathetic ears. The government was favorable to botanical research, and had a flourishing garden at St. Petersburg with specimens of vegetation from every clime. Better still, Emperor Nicholas I was a Fellow of the Horticultural Society of London, Douglas's employer.[3] Best of all, the director of the garden, Dr. Gotthelf Fischer, was a friend of Hooker's; and Hooker's pupil, John Goldie, had been engaged to collect plants for the garden.[4] Here was another opportunity for Hooker to assist Douglas, so he interceded with Fischer and obtained his favorable influence.

The result was that the plan was approved and was referred to Baron Wrangel, Governor of Alaska, apparently with the sanction of the emperor himself—"backed by Imperial favour from the court," Douglas puts it. The Governor, a distinguished nobleman keenly interested in natural science and anxious to assist its laborers, received the proposal in a spirit of the heartiest coöperation and wrote Douglas in the words at the beginning of this chapter. The botanist received the letter while in California, and received a similar letter after his return to the Columbia, Wrangel having feared that the former one might have been lost.

Douglas was greatly pleased with the kind invitation to Russian America and the promise of assistance for his journey home by Siberia. While in California he had mentioned the matter to Cyril Klebnikoff, head of the Russian American Fur Company, and some officers of the Russian Navy who were visiting there, all of whom gave encouragement and extended their kind wishes. Altogether the Muscovites made a very favorable

[3] *Transactions*, Horticultural Society of London, VII (1830), Appendix.
[4] John Goldie, "In Memory of David Douglas," pp. 89-90.

impression on him: "Indeed, they seem to be a set of people whose whole aim is to make you happy."[5]

Under these propitious circumstances he determined to leave the Columbia in the spring of 1833. He would go north to Fort St. James, the Hudson's Bay Company's headquarters in New Caledonia, thence westward to the Russian settlements on the seacoast, where he would obtain the promised transport to Siberia, and then travel across Siberia to Europe and England. He would have plenty of time to reach Sitka, Alaska, by May of 1834, and Wrangel's promise probably would be as good then as it was for May of 1832. In those far-distant parts of the world things happened only about once a year anyway—the Hudson's Bay express, for example. He looked forward to his return to Scotland and hoped to have "a fine jaunt to the Highlands" with Hooker's son, "perhaps in the summer of 1835."

But while the prospects for the Russian part of the journey were encouraging, those for the preceding portion were the opposite. New Caledonia, the territory now comprising the northern interior of British Columbia, had a bad reputation among the fur traders. The climate was severe, the temperature sometimes reaching 55° below zero. Food was scarce, the staff of life being salmon, fresh or dried (chiefly the latter), and starvation was not unknown.[6] The trading posts were isolated and the Indians troublesome. Altogether it was a territory to be avoided. Indeed, the fur traders had such dread of being sent there that they looked on it as another Botany Bay.[7]

Douglas was alarmed by what he heard about it. "The coun-

[5] Encouragement was also given by Captain Fedor Petrovich Lutke of the last Russian voyage of discovery (1826) and his botanist, Karl Heinrich Mertens. Douglas saw them in London and spent some days in going over their botanical discoveries on the Alaskan Coast and Islands. Lutke gave him letters of introduction to Wrangel and "circulars to Siberia."

[6] Morice, *History*, pp. 177-178; P. S. Ogden, "Peter Skene Ogden's Notes on Western Caledonia," W. N. Sage, ed., *British Columbia Historical Quarterly*, I (1937), 52.

[7] John Tod, "New Caledonia," MS., Bancroft Library, p. 52.

try is certainly frightful," he wrote Hooker, "nothing but prodigious mountains to be seen: not a deer comes, say the Indians, save once in a hundred years—the poor natives subsist on a few roots." He was so concerned that he had grave fears for his safe return; and as he was anxious that Hooker should know what he saw and did "on this important journey," he resolved to write for him a daily journal, from which the most important and interesting things could be extracted, whether he survived or not.

Another discouragement was the fever epidemic which still raged. Whites as well as natives were struck down. "Only three individuals out of one hundred and forty altogether escaped it, and I was one of that small number," says Douglas, evidently ignoring his brush with the fever a few years before.

Greatest discouragement of all was his failing eyesight, which was worse than ever. While the left eye had become clearer and keener in its vision, the sight of the right eye had gone entirely and was "as dark as midnight." Looking through a telescope or microscope he could generally see objects at a short distance fairly well, but the least fatigue brought on a doubling of the object, with a surrounding vapory haze that soon concealed everything. This was the result of the ophthalmia and snow-blindness previously mentioned (in Chapter 13), and it had been put beyond recovery by the scorching heat and glare of the sun in California. Purple goggles gave some relief, but he used them "most reluctantly, as every object, plants and all, is thus rendered of the same colour," quite a handicap for a botanist!

Again his determination brushed aside his discouragements; he decided to make a start and go as far as he could, even though he be forced to return. Leaving Fort Vancouver on March 20, 1833, he traveled with the express as far as Fort Okanagan,[8] then set out over territory new to him. Horses were

[8] Barnston, "Abridged Sketch," p. 277.

used to ascend the Okanagan Valley along the east side of the river and the west side of the lake. From the head of the lake a trail led to Fort Kamloops (often called Thompson's River Post), where the horses were changed and the journey continued to Fort Alexandria on the Fraser River. The remainder of the journey was by boat up the Fraser, Nechaco, and Stuart Rivers to Fort St. James on Stuart Lake.[9]

Although he began the journey with some trepidation he was well provided with equipment—a marked difference from his expeditions of earlier years. For food he had rice and biscuit, with tea, coffee, and sugar, and even a gallon of wine. He had a small tent, and his clothing included a straw hat, a pair of deer-skin trousers, and twelve pairs of moccasins, as well as two pairs of shoes. As presents for the Indians he took a supply of Irish roll tobacco, scalping knives, beads, coat buttons, and hair ribbon.[10] Two companions of former journeys went with him: his servant, Johnson, "and my old terrier, a most faithful and now, to judge from his long grey beard, venerable friend, who has guarded me throughout all my journies, and whom, should I live to return, I mean certainly to pension off on four penny-worth of cat's-meat per day."

After a few days at Fort Okanagan, the horse journey with the New Caledonia brigade was begun. They traveled slowly, and Douglas was able to botanize, take astronomical observations, and fix geographical positions along the way. Regularly during the whole of the journey he never allowed twelve hours to pass without taking an observation of the sun or stars.[11] He

[9] For details of the route see A. G. Harvey, "David Douglas in British Columbia," *British Columbia Historical Quarterly,* IV (1940), 228-229, and authorities there cited; also David Douglas, Book of Field Sketches, Fort Okanagan to Quesnel River (1833), MS., British Columbia Archives, Victoria.

[10] Hudson's Bay Company's Transfer Book B 239 CC 2, 1831-1835, fol. 897-901, Hudson's Bay Co. Archives.

[11] Edward Sabine, "Observations," p. 12. For a partial list of Douglas's observations, together with modern survey observations, see A. G. Harvey, "David Douglas in British Columbia," p. 230 and Appendix A.

also made field sketches of the country traveled from Fort Okanagan to the Quesnel River, showing the natural configuration, with notes on trees, vegetation, and soil.[12]

Fort Kamloops, established by the Astorians in 1812, was the halfway house, and here the horses were changed. During the few days' stay Douglas had a serious altercation with Chief Trader Samuel Black (formerly of Walla Walla), who was in charge of the post. The incident is related by the historian H. H. Bancroft as follows:

> For one who had received from the Hudson's Bay Company nothing but kindness, David Douglas was somewhat free with his comments. He did not like to see that powerful organization which was so ready at all times to sacrifice human life on the altar of their own avarice, so cold and selfishly indifferent outside of their money-making to anything affecting the weal or woe of their fellow-creatures. And the shaggy Scotchman was not afraid to tell them so.
>
> Samuel Black was then in command of Fort Kamloop, and thither David Douglas in his wanderings repaired. While enjoying the lonely hospitality of his brother Scot, and discussing the affairs of the company, Douglas, who was more fiery than politic, exclaimed: "The Hudson's Bay Company is simply a mercenary corporation; there is not an officer in it with a soul above a beaver-skin."
>
> Black was up in arms in a moment. He informed his guest that he was a sneaking reprobate, and challenged him to fight. As it was then dark the duel was postponed until next day. Bright and early in the morning Black tapped at the pierced parchment which served as a window to the guest-chamber, and cried out, "Misther Dooglas! are ye ready?" But the man of flowers declined the winning invitation, and saved his life only to yield it not long after in that luckless wild-cattle pit.[13]

Probably the duel was called off because Douglas, on second thought, realized that in his impatience he had been extremely discourteous and unjust to his host. Black, though a fur trader, was not indifferent to science, being fairly well educated [14] and

[12] Douglas, Book of Field Sketches.
[13] Bancroft, *Northwest Coast,* II, 510.
[14] *Ibid.*

a brother of the editor of the *London Morning Chronicle*.[15] Moreover, his enormous stature and powerful physique may have reminded Douglas that discretion was the better part of valor. The gentlemen of the Hudson's Bay Company were not accustomed to criticisms by visitors (Douglas was the first person other than a fur trader or an Indian to travel through the country), and it is easy to imagine how his jibes nettled a man of Black's temperament.

Leaving Fort Kamloops, the brigade went on to Fort Alexandria, where the horses were changed for boats. Douglas, however, evidently continued on horseback or afoot to a point a little beyond the Quesnel River, where his field sketches cease.[16] The strong, steady current of the Upper Fraser was a severe obstacle to the loaded boats and necessitated the constant use of the pole and tracking line.[17] The journey was also upstream on the Nechaco and Stuart Rivers. Douglas went up the Nechaco to Fort Fraser on Fraser Lake, and probably went overland from there to Fort St. James, where he arrived about June 6. Here, at the second permanent settlement in British Columbia,[18] he was the guest of Chief Factor Peter Warren Dease, the Company's superintendent in New Caledonia.

Fort St. James was a disappointment to Douglas, not in itself, but in the unfavorable conditions he found there for continuing his journey. A party was preparing to leave on an exploring expedition down Simpson's (now Skeena)[19] River, and at first he was disposed to accompany them. However, since it was doubtful whether and when they would reach the sea or any Company post there, he decided not to go. He had come an arduous 1150 miles from Fort Vancouver[20] on what he had

[15] A. C. Anderson, "History of the Northwest Coast," MS., Bancroft Library.
[16] David Douglas, Book of Field Sketches.
[17] Anderson, "History," p. 44.
[18] Founded 1806. Fort McLeod, the oldest, was founded in 1805.
[19] Morice, "History," p. 136; *Journal of the Royal Geographic Society,* XI (1841), 250-251.
[20] F. M. Buckland, "The Hudson's Bay Brigade Trail," *Sixth Report,* Okanagan Historical Society (Vancouver, B. C., 1936), pp. 15-16.

hoped was his way home to England. He now found himself in the wilds of New Caledonia about five hundred miles from the nearest Hudson's Bay post on the seacoast (Fort Simpson, at the mouth of the Nass River),[21] with another three hundred miles to the Russian headquarters at Sitka. To reach Sitka meant weeks, if not months, of travel through new country along little-known rivers and waterways and over difficult portages. Moreover the natives were hostile, especially toward the coast, where they were too plentifully supplied with liquor by American trading vessels.[22]

For once Douglas's determined spirit was beaten. Reluctantly he realized that as far as getting to Alaska was concerned he had come all the way for nothing. Indeed, he seemed further off than ever. The only thing to do was to turn back to the Columbia. From there he might get boat passage to Sitka. Accordingly, after a brief stay at Fort St. James, during which he took some observations, he and Johnson embarked in a small birch-bark canoe and descended the Stuart and Nechaco Rivers. At Fort George (now Prince George) they stayed a day or two with George Linton, clerk in charge, then began descending the Fraser.

Now to disappointment was added disaster. On the 13th day of June at the "Stony Islands," as he called the rocky islets in what is now Fort George Canyon, they were wrecked in the rapids, the canoe was smashed to pieces, and they barely escaped with their lives. Douglas was carried down the rapids into the whirlpool below, and was swept about in the turbulent waters for an hour and forty minutes[23] before being washed onto the rocky shore. He lost everything he had except his instruments, his astronomical journal, notes and charts, and his barometrical observations. Food and blankets were all lost, also

[21] Bancroft, *Northwest Coast,* II, 623.
[22] Merk, *Fur Trade,* pp. 301, 309, 311, 314; Bancroft, *Northwest Coast,* II, 693; F. W. Howay, "Liquor Amongst the Indians," *British Columbia Historical Quarterly,* VI (1942), 165-169.
[23] Letter, Nov. 11, 1833, Douglas to W. E. P. Hartnell, in *Madrono,* II, 98.

his botanical notes and a collection of about four hundred species of plants, two hundred and fifty of which were mosses, some of them new. What concerned him most was the loss of the diary he had been writing daily for Hooker, which could not be replaced.

Fort George Canyon, it may be noted, has always been a dangerous place. Before steamboats could get through it a channel had to be blasted out. As recently as 1937 two men descending in a small boat were wrecked there and nearly drowned—an experience similar to that of Douglas and Johnson more than a hundred years before.

Making their way back to Fort George as best they could, Douglas and Johnson obtained another canoe from Linton,[24] and this time descended the Fraser safely to Fort Alexandria. There they obtained horses and proceeded to Fort Kamloops and Fort Okanagan by the same route as that by which they had come. There is no record of any further quarrel with Black. After his recent hardships Douglas no doubt appreciated better the rude comforts afforded by the Company's posts and was not in the mood for criticism.

Getting two Indian canoes at Fort Okanagan, they descended the Columbia to Fort Walla Walla. From there Douglas again visited the Blue Mountains (for the fourth time) and botanized more sedulously than ever, trying to make up for his heavy losses.

He returned to Fort Vancouver in August, worn out in body and broken in spirit. His hopes for the journey home by Siberia, which he had been building up for years, had been shattered in the first stage; following that had come the river disaster. He had suffered severely from hunger and exposure, and had been faced with, "indeed, nearly utter starvation." [25] "I can not detail to you," he wrote Hooker, "the labour and anxiety this occasioned me, both in body and mind, to say nothing of the hard-

[24] Letter, April 15, 1836, Archibald McDonald to W. J. Hooker, Kew Gardens.

[25] Letter, Nov. 11, 1833, Douglas to W. E. P. Hartnell, *supra.*

ships and sufferings I endured. Still I reflect, with pleasure, that no lives were sacrificed. . . . This disastrous occurrence has much broken my strength and spirits." Then, too, besides the loss of his collections, journals, and other belongings, the four best months of the year for travel were gone.

Yet despite his cruel experiences he blamed no one: "Such are the not unfrequent disasters attending such undertakings. On the whole I have been fortunate, for considering the nature and extent of the Country I have passed over (now 8 years here) and the circumstances under which I travelled my accedents have been few." [26]

Nor had he any harsh feelings against New Caledonia. Instead he uttered an encouraging note of prophecy for its future: "The Interior, north of the Columbia, for the space of nearly 4° of Latitude, and 7° of Longitude, is a beautiful and varied country; the soil is generally fertile; and even to a much higher Latitude it is well worth looking after." [27] In a letter to Hooker he made another interesting remark: "The country over which I passed was all mountainous, but most so towards the Western Ocean:—still it will, ere long, be inhabited."

Exaggeration of the distance traveled crept into some of his letters, one stating he had been to 57° north latitude,[28] and another that "I have been in the Snowy Mountains as high as the 60° over a dreary inhospitable country," [29] notwithstanding that his determination of the latitude of Fort St. James was only 54° 26′ 46″ [30] (almost the same as the modern figures of 54° 26′ 34″).

The season of 1833 had been an unfortunate one for him; everything seemed to have gone wrong. Nor did northwest America look promising for the immediate future. Between the

[26] Letter, Sept. 25, 1833, Douglas to W. F. Tolmie, British Columbia Archives, Victoria.

[27] Edward Sabine, *"Observations,"* p. 12

[28] *Ibid.*, p. 13.

[29] Letter, Nov. 11, 1833, Douglas to W. E. P. Hartnell, *supra.*

[30] Edward Sabine, *"Observations."*

fever epidemic and the hostility of the natives, travel northward along the coast was unsafe. Faced with these difficulties, Douglas gave up entirely his idea of the Alaskan-Siberian journey.

The Sandwich Islands beckoned him—the tropical archipelago whose enchanting beauty and opportunities for botanical and astronomical observations had drawn him more and more strongly each time he had visited it. Instead of going home by Siberia he would go by the Islands; so on October 18, 1833, he left the Columbia, "a land where his discoveries had furnished him frequently with the brightest moments of the purest joy, and where also his losses had caused him days of the most poignant sorrow and regret." [31]

While he had been away on his New Caledonian journey there had arrived at Fort Vancouver from England two young doctors, Meredith Gairdner and William Fraser Tolmie. They too had been students under Hooker, and upon his recommendation had been sent out by the Hudson's Bay Company to help Dr. McLoughlin in combating the fever epidemic. Both were interested in botany and nature study and looked forward to devoting their spare time to these pursuits. Their anticipation was shared by Douglas, who hoped they would be able to carry on his still-unfinished work. Writing to Hooker regarding his recent misfortunes, he said: "It reconciles me somewhat to the loss, to reflect that you now have friends in that country, who will probably make up the deficiency. I have given Dr. Gairdner my notes on some more new species of *Pinus*. This gentleman and Mr. Tolmie will have a good deal to contend with. Science has few friends among those who visit the coast of North-West America, solely with a view to gain. Still with such a person as Mr. McLoughlin on the Columbia, they may do a great deal of service to Natural History."

Gairdner contracted tuberculosis and died in a few years.[32]

[31] Barnston, "Abridged Sketch," p. 329.

[32] A. G. Harvey, "Meredith Gairdner: Doctor of Medicine," *British Columbia Historical Quarterly*, IX (1945), 89-111.

Tolmie survived and attained prominent positions in the Hudson's Bay Company and on Vancouver Island. He botanized as a hobby and sent plant specimens to Hooker from time to time, some of which were named after him.[33]

Before leaving the Columbia, Douglas did some more mountaineering. His first intention was that he and Gairdner should ascend Mount St. Helens[34] (9,671 feet), which had recently been in volcanic eruption;[35] but he gave up the idea, probably owing to difficulty in getting the natives to go anywhere near the dreaded phenomenon. He then was drawn to Mount Hood (11,245 feet), whose shining, snow-clad dome had attracted him when he first arrived in the country. He made an attempt to reach the summit but was unsuccessful. It may have been due to lack of proper equipment; or perhaps his experience was similar to that of Tolmie, who attempted to climb Mount Rainier (14,408 feet) at about the same time: Indian guides were used as far as they dared go without offending the Great Spirit of the Mountain, leaving the enthusiastic naturalist to struggle on alone until exhaustion, storm, or sheer acclivity forced a retreat.[36]

It is noteworthy that these young men were the first persons to attack the two great mountains, and that their efforts were almost simultaneous. Tolmie, having heard of Douglas's designs on Mount St. Helens, had proposed that he accompany him on his Mount Rainier expedition,[37] confident that its ascent was practicable,[38] but Douglas was busy with Mount Hood. It was many years before the three summits were reached (St. Helens in 1853; Hood in 1857; Rainier in 1870), and then it was accomplished only by organized and equipped mountaineering parties.

[33] S. F. Tolmie, "My Father: William Fraser Tolmie," *British Columbia Historical Quarterly*, I (1937), 227-240.

[34] Letter, Sept. 20, 1833, W. F. Tolmie to W. J. Hooker, Kew Gardens.

[35] Parker, *Journal*, pp. 337-338.

[36] *Washington Historical Quarterly*, I (1907), 77-81.

[37] Letter, Sept. 20, 1833, W. F. Tolmie to W. J. Hooker, *supra.*

[38] Letter, Dec. 31, 1833, W. F. Tolmie to W. J. Hooker, Kew Gardens.

17

FLAKES OF GOLD

IN NOVEMBER of 1935 there was published in the leading American magazines an advertisement which invited visitors to California. The heading was "MORE CURIOUS FACTS ABOUT SOUTHERN CALIFORNIA," and it began as follows:

First discovery of California gold was made *in England* in 1831. (Found on the roots of trees sent back by a Scotch botanist.)

After news of the discovery of gold in California in 1848 reached England, somebody there remembered that the roots of some of the California pines received from Douglas years before had had small flakes of gold in the clotted earth attached to them.[1]

Douglas knew about the gold. Among the tales that he told on his return to the Columbia was one of his having found enough gold on the roots of plants in California to make a watch seal.[2] He does not appear to have recorded it—probably because he did not deem it of sufficient importance. California gold had been heard of much earlier. About 1820 an English ship captain took home a splendid specimen in quartz rock, which was exhibited at the Royal Institution.[3] Even a shipment of a hundred ounces from Monterey to Boston in 1841 did not cause any great stir.[4]

When its importance was realized there were regrets that

[1] *Quarterly Review,* London, no. 87 (1850), p. 416; Bancroft, *California,* III, 403-404.

[2] Roberts, "Recollections," p. 11; Bancroft, *California,* III, 403-404.

[3] *Quarterly Review,* no. 87 (1850), p. 416.

[4] *Century Magazine,* XLI (1890), 176, 783-784.

something had not been done about it before. Among those blamed were the early naturalists, who were charged with having been so engrossed with the flora and fauna that "they had not an eye to the main chance and saw no gold."[5] But in answer it was pointed out that the men more properly open to censure were those whose especial business had been to notice the treasure, such as Fremont, the explorer, and Dana, the geologist.[6]

One reason the extent of the rich gold deposits was not discovered earlier may have been the opposition of the mission fathers. They are said to have discouraged gold hunting as being demoralizing to the Indians and to have kept secret the existence of the precious metal.[7]

To the north in the Columbia region on the shore of Okanagan Lake, Douglas again found gold enough to make a seal.[8] He passed along the west side of the lake on his journey to and from Fort St. James in 1833 (see Chapter 16). Probably the discovery was made at the mouth of some creek,[9] several of which have produced gold in later years. Once more his letters are silent on the matter. His journal may have mentioned it, but that was lost in the Fraser River. No doubt the information came to the recorder from some Hudson's Bay Company man, as in the case of the California discovery.

Another mineral discovery credited to Douglas is the great deposit of carbonate of lead, galena, and copper situated on the east shore of Kootenay Lake in British Columbia, now known

[5] *Quarterly Review*, no. 87 (1850), p. 416; Bancroft, *California*, III, 403-404.

[6] *Quarterly Review*, no. 87 (1850), pp. 417-418.

[7] *Century Magazine*, XLI (1890), 417; Alfred G. Lock, *Gold—Its Occurrence and Extraction* (London, 1882), pp. 123-124.

[8] W. C. Grant, "Remarks on Vancouver Island," *Journal of the Royal Geographic Society*, XXXI (1861), 213.

[9] T. A. Rickard, "Indian Participation in the Gold Discoveries," *British Columbia Historical Quarterly*, II (1938), 5. One spot Douglas visited was the mouth of Lambly (Bear) Creek, where he took observations (*British Columbia Historical Quarterly*, vol. IV, 1940, p. 242). Mr. F. M. Buckland of Kelowna, on the opposite shore of the lake, has told the author that near this spot about the year 1900 a friend of his washed out gold dust worth $2000.

as the Bluebell Mine.[10] The story has often been retold, but it cannot be true, for Douglas was never on any part of the shore of Kootenay Lake. His carefully written journal, with its daily entries and detailed list of his travels during 1825-1827, makes no mention of it, and in fact leaves no room for it.[11] Nor is it shown on the map of his route in Hooker's *Flora Boreali-Americana.* On his return to the Columbia in 1830 he went only as far as Fort Walla Walla and the Blue Mountains; while in 1832-1833 his travels were on the lower Columbia and to Puget Sound, New Caledonia, and again Walla Walla and the Blue Mountains. Except for his overland journey to Hudson Bay in 1827, during which he passed the mouth of the Kootenay (or McGillivray's) River,[12] his nearest approach to Kootenay Lake was Kettle Falls (Fort Colville). He was there three times in 1826, spending all together seven weeks,[13] and on one occasion walked twenty miles up the Columbia and back. This was the closest he got to Kootenay Lake.

The pieces of rock collected on the Columbia he took all the way home to England and presented to the Geological Society of London.[14] The collections of this society were transferred to the British Museum (Natural History Department) in 1911, but his specimens were not among them. No trace of them can be found at either place, although the Museum has the list of them written by him. It does not mention any from Kootenay Lake.

Possibly while at Kettle Falls, Douglas saw samples of the ore from the Kootenay Lake deposit and expressed some opinion about it; this may have led to his name's being connected with its discovery.

[10] G. M. Dawson, "The Mineral Wealth of British Columbia," in *Annual Report* for 1887-88, Geological and Natural History Survey of Canada (Montreal, 1889), III, 62 R; W. A. Baillie-Grohman, *Fifteen Years' Sport and Life in the Hunting Grounds of Western America and British Columbia* (London, 1900), p. 232. Neither Dawson nor Baillie-Grohman states the source of the story.

[11] Douglas, *Journal,* pp. 254-255.

[12] *Ibid.,* p. 249.

[13] *Ibid.,* pp. 165-169, 174-179, 203-207.

[14] *Transactions,* Geological Society of London, 2nd ser., II (1829), "List of Donation."

18

HAWAII AND ITS VOLCANOES

DOUGLAS left the Columbia for the Sandwich Islands on the Company's brig *Dryad,* Captain Charles Kipling, the route being via San Francisco, where a stop was to be made for water and provisions. At the mouth of the Columbia heavy gales compelled them to lay by in Baker's Bay for eleven days. On October 18 they ventured out and succeeded in crossing the dangerous bar despite a heavy swell. As shipmates Douglas had Chief Factor Duncan Finlayson and John Ball, the latter having tired of both school teaching and farming.[1]

It was a stormy sail down the coast, for this was one of the occasions when the Pacific Ocean belied its name. Ball suffered much from seasickness and ague. Douglas, experienced voyager that he was, said it was "a tempestuous passage" and that the vessel reached San Francisco "nearly a wreck." They entered the Golden Gate on November 4, and, passing the few dilapidated buildings and three or four abandoned cannon constituting the presidio, anchored a mile or so southward near what Ball called "the shore of little valleys and sand hills, all in their natural forest of bushes and trees," which is now the heart of San Francisco.[2] Next day they received an official visit from Governor Figueroa, Victoria's successor, who dined aboard ship.[3] Before leaving port Douglas wrote him (in Spanish)

[1] Ball, *Autobiography,* pp. 99, 100; John Ball, "Across the Continent Seventy Years Ago," extracts from the journal and letters of John Ball compiled by his daughter, Kate N. B. Powers, *Oregon Historical Quarterly,* III (1902), 105.

[2] Ball, *Autobiography,* p. 100.

[3] *Ibid.,* p. 101; *Oregon Historical Quarterly,* III (1902), 105.

sending a list of the latitudes and longitudes of the missions.[4]

The landing place was at Clark's Point (near the intersection of Broadway and Battery Street),[5] but the nearest settlement was the Dolores Mission, consisting of a few small adobe buildings a mile or two away.[6] During his stay Douglas put up his tent "on the Hill of Yerba Buena"[7] (now Telegraph Hill) and used this as headquarters for his travels in the vicinity. He could not go far because the ship was to sail on short notice, and he did nothing in the way of botany. Strong gales thwarted several attempts to beat through the Golden Gate, however, and they were detained much longer than expected. Accompanied by Finlayson, Douglas went one day to Whaler's Harbor (now Sausalito) and landed at the farmhouse of John Read, where he visited the site of an old Indian camp indicated by interesting mounds of marine shells. Read's rowboat, crossing occasionally to Yerba Buena, has been called the first ferry on San Francisco Bay.[8]

He much wanted to visit the Hartnells at Monterey, but, that being out of the question, he wrote a long letter conveying his kind regards, especially mentioning "Dna. Teresa" (Mrs. Hartnell) in remembrance of "the multitude of kindnesses she bestowed on him when under her roof," and hoping to visit them "at no distant period—though not God willing before I see the 'Land o' Cakes.'" He inquired as to increase in the family (a large one, eventually twenty-five children) and the "names of *the* or little strangers," and sent greetings to all his friends, many of whom he mentioned by name.[9] The warm welcome and kind hospitality he had received in California had not been forgotten.

The Dryad left San Francisco on November 29, and after a

[4] Letter, Nov. 25, 1833, Bancroft Library. See list in Bancroft, *California,* III, 404.

[5] *Madrono,* II (1933), 98.

[6] Ball, *Autobiography,* p. 100.

[7] Letter, Nov. 11, 1833, Douglas to W. E. P. Hartnell, in *Madrono,* II (1933), 98.

[8] Bancroft, *California,* V, 689.

[9] Letter, Nov. 11, 1833, Douglas to W. E. P. Hartnell, *supra.*

diversity of weather[10] arrived at Honolulu on December 23. Christmas Day Douglas spent at the home of Consul Charlton, and plans were made for proceeding to the island of Hawaii to visit the great volcanoes. On the 27th he took passage on an American schooner, and, after stopping at Lahaina for ballast and visiting the mission school at Lahainaluna, he landed at Byron's Bay (now Hilo) on January 2. He was made welcome

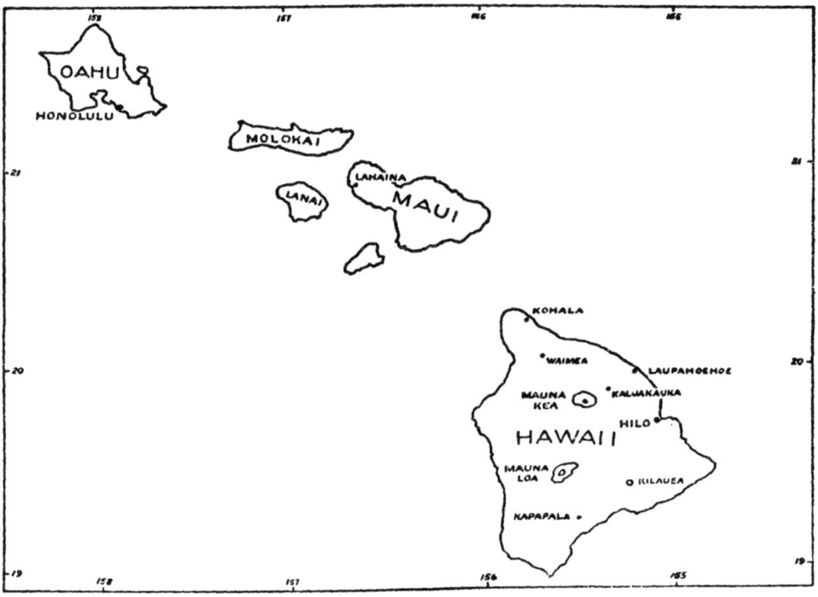

PART OF THE SANDWICH (HAWAIIAN) ISLANDS
(Scale: 1 inch = 65 miles)

by Rev. Joseph Goodrich, pioneer American missionary, and used the Goodrich home as his headquarters during his stay.

Hawaii contains the highest mountains of any island in the world and it has the two largest active volcanoes anywhere on the globe. First to receive Douglas's attention was Mauna Kea (White Mountain), which he had been told was "the highest ground in the known world." Though not entitled to that dis-

[10] Ball, *Autobiography*, pp. 102-103.

tinction, it is the highest *island* mountain and occupies more than half of the northern part of Hawaii. Goodrich had been up it more than once, being interested in geology,[11] so he advised Douglas regarding the route and assisted in the arrangements.

Engaging a party of natives headed by John Honolii, guide and interpreter,[12] he began the twenty-seven mile ascent on January 7. Despite his helpers he did not travel light, but carried on his back a load of more than sixty pounds, consisting chiefly of his instruments. Apparently the route taken was that now generally followed by the Kaumana-Puu Oo trail.[13]

Passing through a cultivated belt about four miles wide where the breadfruit, banana, sugar cane, taro, and other food-bearing plants grew in profusion and luxuriance, they entered a densely wooded area where immense acacia trees (*Acacia koa*) predominated, from which the natives made their canoes. Some of the trees were covered with creeping and hanging plants which, with the undergrowth of tree ferns from four to forty feet high, freshened daily by fog or showers and seeming to "vie with each other in luxury of dress," presented unusually beautiful scenery. But the heavy rain that freshened plants and flowers made travel difficult. The path was so soft that repeatedly they sank in the mud above their knees, and the creeks were high and dangerous to cross.[14] They stopped at a sawmill for dinner, and passed an uncomfortable night at some huts where there was no dry fuel.

Next day the path became worse as the rain continued. However, the vegetation was not overlooked. High up on the mountain Douglas found "a gigantic species of the *Compositae* with

[11] *American Journal of Science and Arts,* XI (1826), 1, 2; XXV (1834), 199-203.

[12] W. F. Wilson, *David Douglas, Botanist, at Hawaii* (Honolulu, 1919), p. 46. Honolii was one of the young men whose visit to the United States led to the sending of the first missionaries to the Islands.

[13] L. W. Bryan, "Kaluakauka," *Paradise of the Pacific, Honolulu,* December, 1934, p. 29.

[14] The Hilo district is notorious for its heavy rainfall, which has been known to reach as much as 386 inches in a year—greater than an average of an inch a day.

a column of imbricated, sharp-pointed leaves, densely covered with a silky clothing." This was the famous silver sword plant which had been discovered by James Macrae, whom the Horticultural Society had sent to the Islands with Lord Byron's party in 1825. Douglas sent home seeds of the plant, and it was named after him by Hooker (*Argyrophyton Douglasii*).[15] The present name is *Argyroxiphium sandwicense*.[16] Wet to the skin and numb with cold, they arrived for the night at a cattle hunter's lodge on the bank of a craggy lava stream up the mountain slope above the forest. On the way Douglas met Messrs. Miles and Castle, owners of the sawmill, who next day went out and shot a young bull for the use of his party. This was one of the many wild cattle which roamed and flourished on the grassy flanks of the mountain. Three days were spent in exploring some of the extinct volcano peaks, examining the vegetation, collecting mosses, and preparing for the summit climb. More food was obtained by shooting another of the wild cattle.

Early on the morning of the 12th Douglas climbed to the summit, accompanied by three Islanders and two Americans (probably Miles and Castle). Arriving soon after ten o'clock, he spent the rest of the day examining the wonders of nature and making instrument observations. The intense dryness of the air peeled the skin off his face and hands as if he were affected with leprosy. His eyes became bloodshot and his eyelids stiff, and a violent headache set in.

From his barometrical observations he estimated the height of Mauna Kea at 13,851 feet,[17] which is surprisingly near to accuracy, the modern determination by the United States Government Survey being 13,784, a difference of only 67 feet. It

[15] W. J. Hooker, *Icones Plantarum* (London, 1837), t. 75; David Douglas, "Mr. Douglas' Voyage from the Columbia to the Sandwich Islands, and the Ascent of Mouna Roa," in W. J. Hooker's "Memoir," *OHQ*, VI, 421.

[16] W. A. Bryan, *Natural History*, pp. 206, 227-229.

[17] Letter, May 6, 1834, Douglas to W. J. Hooker, in W. J. Hooker, "Memoir," *OHQ*, VI, 417. Elsewhere he gives the height as 13,645 feet (*Journal*, Royal Geographic Society, vol. IV, 1834, p. 342).

had been estimated as high as 18,000 feet. Goodrich assisted by making simultaneous barometrical observations at Hilo.

This was the highest mountain Douglas had ascended, and the exhilarating altitude and awe-inspiring surroundings affected him with sensations similar to those he had experienced at the summit of Mount Brown:

> Were the traveller permitted to express the emotions he feels when placed on such an astonishing part of the earth's surface, cold, indeed, must his heart be to the great operations of Nature, and still colder towards Nature's God, by whose wisdom and power such wonderful scenes were created, if he could behold them without deep humility and reverential awe. Man feels himself as nothing—as of standing on the verge of another world. The deathlike stillness of the place, not an animal nor an insect to be seen—far removed from the din and bustle of the world, impresses on his mind with double force the extreme helplessness of his condition, an object of pity and compassion, utterly unworthy to stand in the presence of a great and good, and wise and holy God, and to contemplate the diversified works of his hands! [18]

With the approach of night he made a hasty descent to the lodge, highly gratified with the results of the day's very fatiguing journey. Stormbound there by rain, fog, and snow for two days, they descended the mountain to the sawmill on the 15th and reached the bay next day. The violent rain, slippery path, and swollen streams made traveling difficult and dangerous; nevertheless, he gathered "a truly splendid collection of Ferns, of nearly fifty species, with a few other plants, and some seeds, which were tied up in small bundles, to prevent fermentation, and then protected by fresh Coa bark. Several beautiful species of Mosses and Lichens were also collected; and spite of all the disadvantages and fatigue that I underwent, still the magnificence of the scenery commanded my frequent attention, and I repeatedly sate down, in the course of the day, under some huge spreading Tree fern, which more resembled an

[18] Douglas, "Voyage," *OHQ*, VI, 424.

individual of the Pine than the fern tribe, and contemplated with delight the endless variety of form and structure that adorned the objects around me. On the higher part of the mountain I gathered a Fern identical with the *Asplevium viride* of my own native country, a circumstance which gave me inexpressible pleasure, and recalled to my mind many of the happiest scenes of my early life." [19]

On the way down Honolii was nearly drowned in the river, having to cling to a rock a short distance above a chain of waterfalls until extricated by two men.

Douglas paid his men, excepting Honolii and another ("the King's man"), two dollars each. Some were paid in money and some in goods—cotton, cloth, combs, scissors, thread, etc.—while those especially willing and active were given small presents as well. "Most of them preferred money, especially the lazy fellows." Of the sixteen men, only five were used for carrying Douglas's baggage and provisions, the other eleven being required for carrying their own food and equipment. "Nor was this unreasonable," says Douglas, "for the quantity of *Poe* which a native will consume in a week, nearly equals his own weight! A dreadful drawback on an expedition. Still, though the sixteen persons ate two bullocks in a week, besides, what they carried, a threatened scarcity of food compelled me to return rather sooner than I should have done, in order that the Calabashes might be replenished. No people in the world can cram themselves to such a degree as the Sandwich Islanders; their food is, however, of a very light kind, and easily digested." [20]

Poi (or poe), the national dish and the staff of life of the Islanders, was made from the taro, a species of starchy plant (*Colocasia antiquorum*) growing in damp soil. The roots were thoroughly cooked and then pounded into a pulp, water being added to give it the consistency of paper-hanger's paste. It was served in a calabash (native bowl) and eaten with the fingers—

[19] Douglas, "Voyage," *OHQ*, VI, 425.
[20] Douglas, "Voyage," *OHQ*, VI, 426.

sucked off—and was preferred after fermentation had begun. As it was very fattening, and the natives looked on obesity as beauty, it was gormandized in enormous quantities. To increase their consumption the chiefs and nobility resorted to the *lomi-lomi.* This consisted of stuffing the stomach to the limit of endurance and then reclining while a servant vigorously kneaded it to give relief, this procedure being repeated until the desired result, a sort of stupefied satiety, was obtained.[21]

After spending a few days at Hilo arranging and preserving the plants he had collected, Douglas set out on another expedition on January 22. This was to the world's greatest active volcano, Kilauea. Honolii again accompanied him, this time with nine men. As usual there was a great quantity of food and luggage to be carried, "while each individual," Douglas records, "provided himself with the solace of a staff of sugar cane, which shortens the distance, for the pedestrian, when tired and thirsty, sits down and bites off an inch or two from the end of his staff."

Staying for the night at a native hut, he was amused next morning at seeing an old woman cramming poi into the mouth of a dog that she was fattening to eat. The poi that ran out the side of the dog's mouth she herself picked up and ate. Shortly before daybreak his host went to the door and called out some extraordinary sounds, in response to which came a loud grunt from beneath the tree-ferns nearby. This was followed by the appearance of a fine, large black pig, which was killed and cooked on heated stones for the use of Douglas and his men.

Reaching the summit of Kilauea early in the afternoon, Douglas walked around the west side of the crater, the most active part, and sat down, "not, correctly speaking, to enjoy, but to gaze with wonder and amazement on this terrific sight, which inspired the beholder with a fearful pleasure." The volcano was in a state of comparative tranquillity, but was still an awe-inspiring sight:

[21] J. K. Townsend, *Narrative of a Journey across the Rocky Mountains to the Columbia River, and a Visit to the Sandwich Islands, Chili, etc.* (Philadelphia, 1839), pp. 199-200.

A lake of liquid fire, in extent about a thirteenth part of the whole crater, was boiling with a furious agitation; not constantly, however, for at one time it appeared calm and level, the numerous fiery red streaks on its surface alone attesting its state of ebullition, when again, the red-hot lava would dart upwards and boil with terrific grandeur, spouting to a height which, from the distance at which I stood I calculated to be from forty to seventy feet, when it would dash violently against the black ledge, and then subside again for a few moments. Close by the fire was a chimney about forty feet high, which occasionally discharges its steam, as if all the steam-engines in the world were concentrated in it. This preceded the tranquil state of the lake, which is situated near the Southwest, or smaller end of the crater. In the center of the Great Crater a second lake of fire, of circular form, but smaller dimensions, was boiling with equal intensity; the noise was dreadful beyond all description.[22]

His tent was pitched for the night on the north side of the crater twenty yards back, but his men slept in a ti leaf hut only six feet from the edge. As the sun went down the splendor increased, and "when the nearly full moon rose in a cloudless sky, and shed her silvery brightness on the fiery lake, roaring and boiling in fearful majesty, the spectacle became so commanding, that I lost a fine night for making astronomical observations, by gazing on the volcano, the illumination of which was but little diminished by a thick haze that set in at midnight."

The fearlessness of the men was in marked contrast to the former attitude toward the volcano. In earlier years, because of their dread of the goddess Pele, who was supposed to control the eruptions, the natives did not dare approach the Kilauea crater without making her an offering of the sacred ohelo berries (*Vaccinium sandivicensis,* a huckleberry); and when an eruption took place large numbers of pigs were thrown into the lava stream to appease her anger. In 1790 she was blamed for the destruction of a whole army division, which was caught in a

[22] Douglas, "Voyage," *OHQ,* VI, 428.

sudden blast while passing the volcano.[23] The spell of the dread goddess was broken in 1824, when some eighty natives led by Kapiolani, a royal princess who had become Christian, defiantly marched down into the crater, sang a hymn of praise to the true God, and returned safely.

Next morning was spent in instrument observations and in examining the numerous cracks and fissures emitting steam or sulphurous vapor. One of the fissures was customarily used as a cooking-place by the natives when visiting the volcano, and here some pork from the pet pig, with a fowl and vegetables, wrapped up in banana leaves were cooked to a nicety in twenty-seven minutes. Douglas estimated the height of the volcano above sea level to be 3,873 feet. It had been estimated as high as 10,000 feet. The modern determination is 4,090 feet.

In the afternoon, with three men he descended several hundred feet right into the crater and to the brink of the black ledge of a "vast basin, recently in a state of igneous fusion, now, in cooling, broken up, somewhat in the manner of the Great American lakes when the ice gives way, in some places level in large sheets, elsewhere rolled in tremendous masses, and twisted into a thousand different shapes, sometimes even being filamentose, like fine hair,[24] but all displaying a mighty agency still existing in this immense depository of subterraneous fire." The hollow lava, "faithless beneath his tread," made him feel uncomfortable.

Of all sensation in nature, that produced by earthquakes or volcanic agencies is the most alarming: the strongest nerves are unstrung, and the most courageous mind feels weakened and unhinged,

[23] Douglas gives the year of this disaster wrongly as 1787, and says he was told by the last of the priests of Pele that 5405 men were killed (letter, May 3, 1834, Douglas to Edward Sabine, in *Journal*, Royal Geographic Society, vol. IV, 1834, pp. 335-336). This incredible figure is probably a mistake in understanding his informant, or in recording the statement. The figure in other accounts varies from 80 to 400. (Charles H. Hitchcock, *Hawaii and Its Volcanoes*, Honolulu, 1909, pp. 165-166.)

[24] Called Pele's hair, after the goddess who was supposed to control the eruptions.

when exposed to either. How insignificant are the operations of man's hands, taken in their vastest extent, when compared with the magnitude of the works of God! [25]

The velocity of the lava flow he estimated at nearly three and a quarter miles an hour.

He remained more than two hours in the crater, where the intense heat and the sulphurous fumes gave him a severe headache and other feverish symptoms. He ascended at the southwest end and returned to his tent, having made a complete circuit of the crater. Awaking next morning at 2:00 A.M., he sat gazing on the roaring pit, where three new fires had burst out since the preceding evening. The fumes made Honolii, a victim of asthma, cough incessantly.

Next day he went to Kapapala to make preparations for the ascent of Mauna Loa (Big or Long Mountain). On the way he passed several large caves or tunnels which were used by the natives for various purposes: occasionally as permanent dwellings; more frequently as cool retreats for making cloth from mulberry bark or making canoes; sometimes for goat pens and pig sties, or for rich garden-plots.

The chief at Kapapala had got ready a clean hut for Douglas to stay in, but the surroundings were so repulsive because of the water-filled canoes containing fermenting mulberry bark, the large pig pen, the abundance of fleas, and the din and bustle of the place, that he declined the chief's hospitality—as he had been obliged to do on a similar occasion in North America—and pitched his tent a little distance away, leaving Honolii to explain the matter to the chief as best he could.

Next day, Sunday, was a rest day—for all except Honolii. To his duties as "guide, interpreter, purveyor, and, I may say, friend (for in every department of his omnifarious capacity he is a good sort of fellow)" was now added that of preacher. Both morning and evening he expounded to and exhorted in their

[25] Douglas, "Voyage," *OHQ*, VI, 430.

own language a congregation of nearly two hundred natives, who were called to the schoolhouse "by the sound of a conch-shell blown by a little imp of a lad, perched on a block of lava, in front." Douglas did not attend, but he went in the afternoon and found the people assembled for scripture-reading in unison under the leadership of the chief, Honolii having withdrawn to prepare his evening sermon. The schoolmaster, a thirty-year-old hunchback about three feet tall, with long legs and a cast in one eye, prompted and corrected them when necessary, his powerful, trumpet-like voice "resembling the voice of a person bawling into a cask."

On Tuesday, January 28, the journey up Mauna Loa, distinguished as the world's largest single mountain mass, was begun. In addition to his Kilauea party Douglas had two special guides, one of them a bird-catcher, whose occupation was the gathering of feathers for the beautiful necklaces (leis) and cloaks of the native nobility. As usual, the men were laden chiefly with their own food: tara roots, poi, both in paste form and dried, sweet potatoes, and the parts of a large goat which Douglas had bought—each man carrying a pole on his shoulder with a bundle hanging from both ends. Honolii brought up the rear carrying a telescope and his umbrella, the latter a constant article of his equipage, be the weather foul or fair. Douglas spent the night beside a camp fire, while the natives "swarmed together in a small cave, which they literally converted into an oven by the immense fire they kindled in it."

Next day he made the summit climb, setting out early with the bird-catcher, a man named Calipio, and two others. The way was extremely difficult, being over rough ledges of lava, some smooth and glassy, while others were lifted into great mounds or sunken into deep valleys, or rent into fissures, ridges, and clefts.

Walking was rendered dangerous by the multitude of fissures, many of which are but slightly covered with a thin crust, and everywhere our progress was exceedingly labourious and fatiguing. As we continued to ascend, the cold and fatigue disheartened the Islanders, who required all the encouragement I could give to induce them to proceed. As I took the lead, it was needful for me to look behind me continually, for when once out of sight, they would pop themselves down and neither rise nor answer to my call.[26]

Toward the summit progress was hindered by deep, heavy snow which had been softened by the sun.

Douglas reached the summit with only Calipio, the bird-catcher and the other two natives all having fallen by the wayside. Kauka (Doctor), by which title they called him,[27] was too fast for them. The summit was a surprise in that its flatness and extent did not permit a view of any other part of the island, not even of Mauna Kea nor the distant horizon, although it was a beautifully clear day. Douglas tried to descend into the crater, but was unable to because of the treacherous fissures, concealed but not sufficiently protected by the snow.

Varying estimates of the height of the mountain had been made by other travelers, one (Captain Cook's) being as much as 18,400 feet.[28] From observations made with his instruments Douglas calculated the height at 13,517 feet.[29] The modern determination by the United States Government Survey is 13,-680 feet, a difference of 163 feet, which may be due to Douglas's having measured at the east side of the crater and the government's at the west side.[30] Near the crater lodged in the hollows of the rocks he found a peculiar liquid of which he tried to bring away a sample, but it destroyed the glass bottle containing it. It has been suggested that it was fluoric acid.[31]

[26] Douglas, "Voyage," *OHQ*, VI, 436.

[27] Letter, April 13, 1936, L. W. Bryan, Hilo, Hawaii, to the author.

[28] James D. Dana, "Geology," vol. X of *United States Exploring Expedition . . . under . . . Charles Wilkes, U. S. N.* (Philadelphia, 1849), p. 156.

[29] Letter, May 6, 1834, Douglas to W. J. Hooker, in W. J. Hooker, "Memoir," *OHS*, VI, 418. Elsewhere he gives the height as 13,230 feet (*Journal of the Royal Geographic Society*, IV, 1834, p. 343).

[30] W. F. Wilson, *David Douglas*, p. 62.

[31] Bennett, *Narrative*, I, 280-281. In a letter to the author written Aug. 8, 1939, Dr. T. A. Jaggar, Research Associate in Volcanology, University of Ha-

The crater's circumference he estimated at six and a quarter miles, with the ancient crater at twenty-four miles, and the depth of the crater ledge 1270 feet. Although in eruption only a year or so previously, the volcano was not now active, and he found the "gigantic mass of slag, scoriae, and ashes" a dreary though terrific scene. He noticed a peculiar musical sound coming from the cracks and fissures which was faint, like musical glasses, but hissed as if it were a swarm of bees. He could not tell whether it came from the escape of internal heat or from the heating of the external air on the surface of the rocks.

Beginning the descent late in the afternoon, he was caught in the dark and forced to shelter in a cave. His perspiration-soaked clothing and the absence of fuel made the cold so uncomfortable, however, that when the moon rose he continued the descent. "Of necessity we walked slowly, stepping cautiously from ledge to ledge, but still having exercise enough to excite a genial heat. The splendid constellation of Orion, which had so often attracted my admiring gaze in my own native land, and which had shortly passed the meridian, was my guide." Reaching some stunted shrubs at two o'clock, they made a fire for the remainder of the night. In the heat and light of the fire Douglas's eyes became violently inflamed and painful from the preceding day's sun glare on the snow, until blood discharged from them, giving some relief. Having had no food or water since early morning he suffered severely from thirst. "Still I slept for a few hours, dreaming the while of gurgling cascades, overhung with sparkling rainbows, of which the dewy spray moistened my whole body, while my lips were all the time glued

waii, says: "The story of David Douglas finding hydrofluoric acid on top of Mauna Loa sounds fantastic. There is pure snow water and ground ice. Fluorine in the condensed water of the gas-collecting tubes of Day and Shepherd 1912 (Bull. Geol. Soc. Amer. Dec. 16, 1913, Vol. 24, p. 592) was in about double the quantity of chlorine, 0.565 gram fluorine in 100 cubic centimeters of water. This was the gas pumped from behind flames of an active lava dome. There was also a trace of fluorine in the Post Card Creek solfatara in Kilauea crater. Glass bottles became etched. There is much fluorine in outgassed rocks of volcanoes. (Shepherd, Amer. Jour. Sci. V, XXXV-A, 1938, p. 326.)"

together with thirst and my parched tongue almost rattled in my mouth."

In the morning, relieving his eyes with a few drops of opium, he continued the descent and arrived at Kapapala at four o'clock. "Gratified though one may be at witnessing the wonderful works of God in such a place as the summit of this mountain presents, still it is with thankfulness that we again approach a climate more congenial to our natures, and welcome the habitations of our fellowmen, where we are refreshed with the scent of vegetation, and soothed by the melody of birds." [32]

Douglas and his man Calipio were the second party to reach the summit of Mauna Loa. The first ascent had been by his friend Menzies forty years earlier.[33] He had estimated the height even more accurately than Douglas, his figure of 13,634 feet being only forty-six feet out. Douglas found that the natives remembered Surgeon Menzies's ascent and that they knew him as "the red-faced man, who cut off the limbs of men, and gathered grass."

Tarrying again at Kilauea on February 4, Douglas made a barometrical measurement of the depth of the crater, computing it to be 1096 feet. He returned to Hilo on the 7th,[34] the first white man (and probably the first man of any kind) to climb to the summits of both the great island peaks, Mauna Kea and Mauna Loa. He had obtained "a most magnificent collection of lava specimens, showing the successive formations from the sea to the summit, besides a princely collection of plants, to show the verdure at different heights." He looked forward to seeing descriptions of the plants published by Hooker in some form or other. Hooker received them and intended to publish a paper on them,[35] but apparently did not.

[32] Douglas, "Voyage," *OHQ,* VI, 440.

[33] Archibald Menzies, *Hawaii Nei 128 Years Ago,* W. F. Wilson, ed. (Honolulu, 1920), p. 199; W. F. Wilson, *David Douglas,* pp. 62-63; Charles H. Hitchcock, *Hawaii and Its Volcanoes* (Honolulu, 1909), pp. 63-79. This contains Menzies's journal of his ascent of Mauna Loa.

[34] *Hawiian Spectator,* Honolulu (April, 1838), p. 101.

[35] W. J. Hooker, "Memoir," *OHQ,* VI, 449.

Douglas remained on the island for some weeks. One afternoon (Feb. 19), while making a series of observations for the purpose of ascertaining whether any diurnal fluctuation existed in the magnetic action, such as in that of the barometer, his observations suddenly became quite irregular. The cause appeared not long afterwards in a violent earthquake that lasted thirteen seconds. The sea receded about a mile for a few minutes, and part of the Kilauea volcano fell in. All night long the earth continued to be agitated and "Mr. Goodrich's house, of wood, rolled like a ship in a storm, but did not fall." The quake greatly affected Douglas's sensitive nature, leading him to remark in a letter to a friend: "The kinder feelings of our nature are blunted by the falseness of friendship, *but how are our feelings roused when we find our parent earth faithless to our steps!* Of all sensations, those produced by earthquakes, in countries agitated by volcanic fires, are the strongest man can feel."[36]

Douglas wrote a detailed account of his ascents of Mauna Kea, Kilauea, and Mauna Loa, but it stops with his return to Kapapala. About his doings of the next few months little information exists. A good deal of time was devoted to ferns, of which he collected two thousand species, ninety of them new.[37] Among them was the giant pala fern, the starchy and mucilaginous rootstock of which when cooked formed a staple native food. Three names, all honoring Douglas, have been given it: *Gymnotheca Douglasii, Marattia Douglasii,* and *Stibasia Douglasii.*

Another plant named after him is the picturesque pandanus or screwpine (*Pandanus Douglasii*). Called *puhala* by the natives, it is noted for its stilt-like aerial roots and the perfect spiral arrangement of its long, sword-like leaves. The latter (*lauhala*) are used to make native mats, fans, etc., while the bright orange-red fruit is used for making leis.

[36] *Sandwich Island Gazette,* Honolulu, July 30, 1836.

[37] A few of these have been traced at Kew Gardens. Others are among Lindley's plants at Cambridge University, according to Gilmour and Tutin's list, but they have not been traced. None have been found at the British Museum.

Douglas left Hilo on March 29[38] and reached Honolulu early in April. Hoping to get passage to England, he waited there until early in July. He occupied the time with something he had long intended: the classification of the plants of Hawaii in zones according to elevations above the sea.[39]

During his travels in America Douglas had seen many wonderful sights and had many thrilling experiences, often accompanied by grueling hardships. Indeed, so accustomed had he become to this combination of wonders, thrills, and hardships, that it seemed part and parcel of his existence. Nor were the events of the past few weeks in Hawaii of a different kind. They had been even more wonderful, more thrilling, and, if that were possible, more difficult. They constituted the climax of all his many experiences; he himself said so.

But they left him exhausted in both body and mind. Twice within a month he had climbed from sea level to altitudes exceeding 13,000 feet. Poorly clothed and accoutred, he had gone from the heat of a tropical climate to the cold rigors of inhospitable alpine regions, his equipment consisting chiefly of heavy scientific instruments! With his usual determination he had driven himself at such a pace that his guides could not keep up with him.

We have seen how on previous occasions he quickly recovered from hardships and sufferings. Not so, this time. He was terribly exhausted and unstrung. He was tormented with rheumatism. Even his mind was affected. In some of his letters he became confused and gave distorted and inaccurate accounts of his recent experiences. An example is a letter to Mrs. Charlton, the consul's wife,[40] written on February 7, the day he re-

[38] Mrs. D. B. Lyman, Journal, MS., Hawaiian Mission Children's Library, Honolulu.

[39] Edward Sabine, "Observations," pp. 13-14.

[40] W. F. Wilson, *David Douglas*, pp. 58-60.

turned to Hilo from Mauna Loa and Kilauea. It is quoted at length[41] to show the distraught state of his mind:

As I have just about an hour ago arrived from Mauna Loa, the volcano, &c., and having enjoyed a bath and an excellent cup of tea, with a willing pleasure I sit down to tell you the story of a traveller . . .

A sight of the volcano fills the mind with awe—a vast basin in a state of igneous fusion, throwing out lava in a thousand forms, from tortuous masses like large cables to the finest filamentous thread. Some places in large sheets, some in terrible rolled masses, like the breaking up of a large river with ice—of all colors and forms, showing the mighty agency ever existing in its immense laboratory. The strongest man is unstrung; the most courageous heart is daunted, in approaching this place. How insignificant are the works of man in their greatest magnitude and perfection, compared with such a place. I have exhausted both body and mind, examining, measuring, and performing various experiments, and *now, I learn that I know nothing;* but this much I know, that volcanoes are the irregular, secondary results of great masses of matter obeying the primary laws of atomic action—that they differ in their intensity—are interrupted in their period, and are aggravated or constrained by an endless number of causes, external and purely mechanical. Of all modes of material combination, volcanoes are the most complicated; to assume, then, that volcanic forces have not been called into action at all times in the natural history of our globe, but also, that in each period they have acted with equal intensity, seems to be a merely gratuitous hypothesis, unfounded on any of the great analogies of nature, and I believe also, unsupported by the direct evidence of fact. This confounds the immutable and primary laws of matter, with the mutable results arising from their irregular combination. It assumes that in the laboratory of nature, no elements have ever been brought together, which we ourselves have not seen combined; and that no forces have been developed by their combination, of which we have not witnessed the effects. And what does all this amount to, but to limit the riches of the kingdoms of nature by the poverty of our own knowledge, and to surrender ourselves to the mischievous, but not uncommon philosophical scepticism, which makes us deny

[41] From the *Hawaiian Spectator*, Honolulu, April, 1838, pp. 101-103.

the reality of what we have not seen, and doubt the truth of what we do not comprehend.

I assure you, Madam, that these islands offer rewards to the naturalist, over all others. I rejoice, in common with others, to know that man seems no longer to be a worshiper at the portal of nature's temple, but is allowed to pass within, and to be so far a partaker of her mysteries, as to see with his intellectual eye the past revolutions of our globe, and to offer reasonable speculations on the future. The dogmas of geology find here a stumbling-block of which it is needless to speak.

I regret having no companion—the soothing words of a friend stimulate us to exertion, and make our labor light. Labor I will; but if I thought that the imagination, the feelings, the high capacities of our nature, were blunted or impaired by such, I should then regard it as little better than a moral sepulchre, in which, like the strong man, I was burying myself in view of my own destiny. But I believe too firmly in the inscrutable attributes of that Being, in whom all truth, of whatever kind, finds its proper lasting place, to think what we in our ignorance call the discordances of nature, will be in everlasting collision. All will be reconciled, and we shall see no longer as "through a glass, darkly," the infinitude, the beauty, the harmony of nature. I must return to the volcano, if it is only to look—to look and admire.

I should give you a note on *Mauna Kea,* but the time will not permit. I shall tell you verbally. Suffice it to say that I reached the culminant point after immense labor, fatigue, anxiety, and some degree of danger. The cold was intense. You may pledge my name for saying that the *Great Crater* is on the very summit of Mauna Kea, at present in an active state. One day there, madam, is worth one year of common existence. This is twenty-seven miles round, and 1274 feet deep. I rested from 12M. till 12 at night, on the mountain, when the wane moon presented herself in silvery brightness, reflecting a glare on the ragged lava like Gothic turrets. With thankfulness and joy, the beautiful constellation of Orion being my guide, I rose to descend to a climate more congenial to my nature, and the habitations of men, the land of flowers, and the melody of birds.

While the first part of this letter quite clearly refers to Kilauea, it is not plain whether the last part refers to Mauna Kea or Mauna Loa. Mauna Kea is named but the measurements refer to Mauna Loa. In neither case had he seen the crater "at present in an active state."

In a letter to Hooker written on May 6 (after his return to Honolulu) he states that Mauna Loa is "at present in terrific activity." Then follows a description containing the terms "liquid fire," "blazing waves," and "fiery fluid," which, with the rate of flow of the lava stream (three and a quarter miles an hour) would be quite applicable to Kilauea, but the depth mentioned (1272 feet) refers to Mauna Loa.[42]

Both letters were published not long after his death, and criticisms of his inaccuracies soon followed, first in Honolulu,[43] afterwards in the United States.[44] Later on it was suggested that, inasmuch as he had written a very reasonable letter to Captain Sabine only five days before the letter to Hooker, his lapse was due to "temporary hallucination" through having dined too well.[45] The proper explanation, however, seems to lie in the state of his health. Apparently he was so worn out physically and so unstrung mentally that he did not realize what he was saying.

It may be added that although Mauna Loa was not astir at the time of his visit it has turned out to be a very active volcano. Eruptions have averaged about one every four years since 1832, and during the past hundred years it has poured out more lava than any other volcano in the world.

[42] It is said that Douglas also reported verbally that he had discovered an active volcano at the summit of Mauna Loa. Bennett, *Narrative,* I, 280-281.

[43] *The Polynesian,* Honolulu, August, 15, 1840.

[44] Dana, *Geology,* pp. 208-209; *American Journal of Science and Arts,* 2nd ser., X (1850), 236-237; James J. Jarves, *Scenes and Scenery in the Sandwich Islands* (Boston, 1844); pp. 252-258.

[45] James D. Dana, *Characteristics of Volcanoes* (New York, 1890), pp. 57-58.

19

LAST JOURNEY

WHILE WAITING in Honolulu for a ship for England, Douglas met Rev. John Diell, chaplain of the Seamen's Mission. A newcomer from the United States, Diell was interested in Douglas's account of the thrilling scenes in Hawaii and decided to see something of them for himself. There being no prospect of a ship for some weeks, and having recovered from his rheumatism, Douglas saw an opportunity for more work on the volcanoes and offered to accompany the missionary on a visit to Kilauea.

Accordingly on July 3, 1834, Douglas and Diell, with the latter's wife, child, and black man-servant (John), left Honolulu in the schooner *Minerva.* At Lahaina, Diell and his family left the ship in order to visit the island of Molokai, agreeing to meet Douglas at Hawaii. Somehow John missed his master's boat and continued his passage with Douglas.

The vessel was detained at Kohala Point at the north end of Hawaii, so Douglas and John landed there on July 9, intending to travel over the mountain (Mauna Kea) to Hilo. A trail led over the northern and eastern slopes, reaching 6000 feet elevation (in about the same location as the present Waimea-Laumaia road). While longer than the trail along the coast, it avoided its wet ground and steep gulches.[1] The distance by this longer, meandering old trail was probably between ninety and a hundred miles.

Douglas and John started off together, but John became lame and was obliged to stop, leaving Douglas to continue alone with

[1] L. W. Bryan, "Kaluakauka," p. 29.

his little terrier, Billy.[2] After staying the night of July 11 at the house of one Davis[3] (Nedo Ranch),[4] early next morning Douglas called at the hut of Edward (Ned) Gurney, an English cattle hunter, who lived in Hamakua at the intersection of the ancient trails running up Kaala at Lahohinu on the mountain slope above Laupahoehoe[5] (near the present Waimea-Laumaia road in the land of Kukaiau, near the Kapoholimuele Gulch).[6] He had breakfast and asked his way. He was told that some natives would soon be going in his direction, but he was too impatient to wait for them. Gurney, therefore, went with him a mile or so, and on parting warned him that about two and a half miles farther on, near a pond, were three bullock pits (large holes covered with branches, grass, and earth, for trapping the wild cattle that frequented the grassy flanks of the mountain). Two of them lay directly in his path, the first being so disguised that without a good deal of caution he would mistake it for the path and fall in. The second and third were open, with a bullock in each. Douglas tried to engage a native to help carry his bundle and instruments but was unsuccessful, so he trudged on, hoping to reach Hilo that evening.

Two or three hours later (about ten o'clock) two natives were passing the bullock pits, when near the third pit one of them noticed a torn piece of cloth. On looking into the pit they saw a trapped bullock, and underneath it the foot and shoulder of a man, the rest of the body being covered with earth from the caving in of the pit. It was Kauka (Doctor); he had been gored and trampled to death by the bullock.

They hurried with the news to Gurney, who returned and shot the animal and extricated Douglas's body. He then looked

[2] Name in letter, Nov. 7, 1834, Gairdner to W. J. Hooker, Kew Gardens.
[3] Chester S. Lyman, *Around the Horn to the Sandwich Islands and California, 1845-1850* (New Haven, 1924), p. 124.
[4] Henry M. Lyman, *Hawaiian Yesterdays* (Chicago, 1906), p. 31.
[5] Eugene Burns, "The David Douglas Mystery," *Honolulu Star-Bulletin*, April 23, 1938.
[6] L. W. Bryan, "Kaluakauka," p. 29.

for the dog and bundle. After a few moments' search the dog was heard barking, and was found guarding the bundle a little way along the path toward Hilo. Scrutiny of Douglas's footprints showed that after stopping to look inside the first and second pits he had proceeded to where the dog and bundle were found. He then returned to the third pit and while looking into it stepped too close and fell in. There, entirely within the power of the trapped and infuriated animal, he suffered a horrible death. Such, at least, was the account given by Gurney.[7]

Wrapping the body in a bullock hide, Gurney engaged the natives to carry it to the seashore, and paid them with four bullocks, which he shot forthwith. The twenty-seven-mile journey took them the remainder of that day and all the next. Gurney went with them, and at Laupahoehoe he got another native to take the body to Hilo by canoe. Leaving on the evening of July 13, the native paddled all night and reached Hilo early next morning.

Diell had arrived there two days before, and he and Goodrich were awaiting Douglas when the native brought the news. Shocked and incredulous, they hastened down to the canoe. As they lifted up the bullock hide which, having served as a stretcher for the journey down the mountain, now served as a shroud, a horrible sight met their eyes.

"Can it be the man with whom we parted but a few days ago," they exclaimed, "and who was then borne up with so high spirits and expectations and whom but an hour previously we were fondly waiting to welcome to our little circle? The answer," continued their report to Consul Charlton, "was but too faithfully contained in the familiar articles of dress—in the features, and in the noble person before us. They were those of our friend. The

[7] A slightly different version is that the attention of the natives was drawn by seeing the dog sitting on Douglas's coat (letter, Nov. 7, 1834, Gairdner to W. J. Hooker, *supra*). Still another version is that they found the dog running around the pit and barking (Stephen Reynolds, Journal, MS., Peabody Museum, Salem, Mass., vol. IV, July 19, 1834).

body, clothes, etc., appeared to be in the same state they were in when taken from the pit: the face was covered with dirt, the hair filled with blood and dust; the coat, pantaloons, and shirt considerably torn." Douglas's watch was broken but still going.

They carried the body to Rev. Sheldon Dibble's house to prepare it for burial. On washing it they found a shocking condition: badly gashed head, broken cheekbone and ribs, terrible bruises. A burial place was chosen under a breadfruit tree in Goodrich's premises. A man was engaged to dig the grave and another to make the coffin.

While digging, Hall, himself an American bullock-hunter,[8] expressed doubt that the wounds had been inflicted by a bullock, and suggested that Douglas might have been murdered. Goodrich had similar doubts. How was it that Douglas had been left to proceed alone without a guide? What had become of John? How was it that Douglas fell into a pit which he already had safely passed? How could he fail to see the bullock or the hole in the cover?

After some discussion the missionaries thought that the opinion of a medical man ought to be obtained as to the cause of the injuries. That meant sending the body to Honolulu, and since there was no regular boat, they had to do something toward preserving the body. Accordingly, they did the best they could with salt. The body was disemboweled, the cavity filled with salt, the coffin stuffed with salt, and coffin and all enclosed in a box of brine.

Before taking the heavy case to the wharf the missionary families and several other foreigners assembled to pay their respects to the memory of Douglas. The gist of the remarks is probably contained in this extract from the journal of the wife of one of the missionaries: "Poor man he has endured incredible hardships by land and sea and suffered every privation for the benefit of the world. He has lost his life in the cause in which

[8] C. S. Lyman, *Around the Horn,* pp. 123, 128.

he was engaged. He has a widowed mother in Scotland to mourn his loss."[9]

Goodrich and Diell wrote a long letter to Consul Charlton giving details of the tragic affair. They were suspicious of Gurney, pointing out that he was an escaped convict from Botany Bay. They also sent two white men (Hall was one)[10] to the bullock pits to investigate. Gurney arrived at Hilo next day bringing Douglas's dog, money, and other belongings, but, although their minds were "greatly relieved" by his narrative of "the probable way" in which Douglas met death, they were not entirely convinced, conceding only that "perhaps" it was a satisfactory account.

The investigation by Hall and his companion was obstructed by a heavy rain which had obliterated Douglas's footprints. Hall's doubt was increased, however, by hearing from Davis, at whose house Douglas spent his last night, that he had seen Douglas have a large purse of money which he thought was gold, whereas nothing of consequence was found after his death. Hall examined the horns of the bullock and found them blunt. He thought it impossible that the deep gashes on Douglas's head could have been made either by horns or by hoofs, and he concluded that Douglas had been murdered by Gurney. So convinced was he that he obtained the whole head of the animal and took it to Hilo to show to the missionaries.[11] It was sent on to Honolulu,[12] but whether in time for the post mortem is not known.

Two other white men appear to have investigated and come to the opposite conclusion. Mrs. Lyman wrote in her journal on August 3, 1834: "Since the death of Mr. Douglas two foreigners have visited the spot where he was killed. They found every-

[9] Mrs. D. B. Lyman, Journal.
[10] C. S. Lyman, *Around the Horn,* p. 123.
[11] C. S. Lyman, *Around the Horn,* pp. 123-124.
[12] Wilkes, *Narrative,* IV, 54.

thing as the man described it, and are satisfied that his statement was correct. This is a relief to our minds." [13]

Not until August 3 did Consul Charlton receive the body. He found it in "a most offensive state." He had it examined by Doctors Gerritt P. Judd and T. C. R. Rooke,[14] who gave the opinion that the wounds had been inflicted by the bullock. It was also examined by the two doctors of H. M. S. *Challenger*, who found nothing to the contrary.[15] Nevertheless, many people were not convinced and adhered to the murder theory.[16]

The funeral was held next day, attended by the foreign residents of Honolulu and by the captain and several officers of the *Challenger*. One of the officers read the service. Burial took place in the burying ground of the native church, which was used for the interment of both natives and foreigners.

The dog (said to have been brought by Douglas all the way from Scotland)[17] was sent back to England in charge of Captain Peter Corney of the brig *Eagle* and given to James Bandinel, clerk in the British Foreign Office.[18]

[13] Mrs. D. B. Lyman, Journal.

[14] Wilkes, *Narrative*, IV, 205.

[15] Stephen Reynolds, Journal, IV, Aug. 3, 1834.

[16] For further consideration of the question see Appendix A.

[17] Peter Parley (pseud. of Samuel G. Goodrich), *Tales about Plants* (London, 1839), p. 37.

[18] Copy of letter, Aug. 10, 1834, T. C. R. Rooke to George Bentham, Kew Gardens.

20

IN TRIBUTE

WORD of Douglas's death reached his relatives through the press.[1] While reading the Liverpool *Mercury* of February 27, 1835, his elder brother, John, came upon a column of miscellaneous news headed "Multum in Parvo" and found this item:

> Accounts from the Sandwich Islands mention that Dr. Douglass, a scientific Englishman, was killed, while on an excursion in the mountains of Hawaii, by a wild bull.

The report had appeared first in the London *Morning Herald* of the day before. Although regarded with incredulity, it was confirmed in a few days by letters received by the Colonial Office[2] and the Horticultural Society.[3]

It caused widespread regret. Learned scientists, humble gardeners, and nature-lovers generally all mourned the premature death of one who had done so much for botany and horticulture. So many rare and beautiful plants had been introduced through his untiring efforts that, as one writer put it, "if we only imagine the British gardens deprived of the plants introduced by Douglas, we shall find them but little further advanced in point of ornamental production than they were a century ago."[4] Hooker and Captain Sabine particularly were

[1] W. J. Hooker, "Memoir," *OHQ*, VI, 441. Name of the newspaper is given incorrectly as the *Liverpool Monthly* in the *Oregon Historical Quarterly* printing of Hooker's "Memoir."

[2] *Ibid.*

[3] George Bentham, Diary, MS., Kew Gardens.

[4] *Gardener's Magazine*, XII (1836), 606.

saddened; they had been planning a great welcome home for him.[5]

On the Columbia, where the news had been received in November, there were similar feelings. "By this catastrophe," wrote Gairdner, "science has lost one of its votaries. . . . I have to look back upon the hours I spent in his society as among my happiest since leaving England's shores." [6] Archibald McDonald wrote that Douglas had shown "indefatigable perseverance in almost every species of danger, privation and hardship for the advancement of knowledge." [7] Barnston, who had known him as "one of the heartiest, happiest mortals in our little society at Fort Vancouver," declared afterwards: "Through eleven years of toil and hardship, danger, difficulty, and loss, he had strenuously worked his way, following his vocation with a devoted spirit and undaunted courage. He feared God, was beloved by his friends, and esteemed by all who ever had the happiness of knowing him. . . . His worth as a useful member of society, and practical botanist, will live and be felt while the study and cultivation of plants and flowers remain pleasing and beneficial to mankind." [8]

"Everybody who had known him lamented him," said Ranald MacDonald in his memoirs, "for with all his enthusiasm in his pursuit, he was ever the most sociable, kindly and endearing of men. A sturdy little Scot; handsome rather; with head and face of fine Grecian mould; of winning address, genial, and with all, the most sincerely pious of men." [9]

It was not long before a plan was brought forward for the erection of a memorial; and, as was fitting, it originated in Douglas's native county. The Perthshire Royal Horticultural Society resolved to put up a monument at Scone and made a

[5] Letters, Nov. 26 and Dec. 13, 1834, Edward Sabine to W. J. Hooker; letter, April 15, 1836, Archibald McDonald to W. J. Hooker, Kew Gardens.

[6] Letter to W. J. Hooker, Nov. 7, 1834, Kew Gardens.

[7] Letter to W. J. Hooker, April 15, 1836, *supra.*

[8] Barnston, "Abridged Sketch," p. 349.

[9] Ranald MacDonald, *Narrative,* p. 137.

world-wide appeal for contributions.[10] After considerable delay an imposing monument twenty-three feet tall was erected in 1841 in the Scone churchyard.[11] On the front is inscribed:

1841

Erected by the lovers of botany in Europe in memory of David Douglas, a native of this parish, who, from an ardent love of science and a desire to promote the improvement in botany, visited the unexplored regions on the banks of the Columbia and southward to California, whence he transmitted a great variety of the seeds of valuable trees and flowering plants, adapted to the climate of Great Britain, and who, after devoting ten years of the prime of life in adding to the Arboretum and Flora of Europe, suffered an accidental and lamented death in one of the Sandwich Islands, on the 12th July, 1834, in the 35th year of his age. Endowed with an acute and vigorous mind, which he improved by diligent study, this eminent botanist uniformly exemplified in his conduct those Christian virtues which invested his character with a higher and more imperishable distinction than he justly acquired by his well-earned reputation for scientific knowledge. A dutiful son, a kind and affectionate brother, a sincere friend, he secured by the rectitude of his moral and religious principles, not less than by the benevolence of his disposition, the esteemed regard of all who knew his worth.

On the back is a long list of some of the trees, shrubs, and plants introduced by Douglas.[12] The monument is still in a fair state of preservation.

[10] *Gardener's Magazine,* XII (1836), 385, 602-609.

[11] A proposal that the British government place tablets at Scone and Honolulu came to nothing. Nor did the government pay anything for the astronomical and magnetic observations made by Douglas. Captain Sabine made a long report on them (see summary in *Transactions,* Royal Society, vol. III, 1830-1837, pp. 471-472), but evaded the question of compensation because he was annoyed at Douglas's relatives for withholding his Sandwich Islands Journal. The only money realized from his estate was £50 for animal skins and birds sold to the Zoological Society, and an unknown sum for his Sandwich Islands herbarium sold to the Horticultural Society. In 1856 the latter society sold all his herbaria, 1460 specimens, to the British Museum. Others of his specimens are at Kew Gardens and Cambridge University, some under his own name, and some in the collections of Hooker and Bentham at Kew and of Lindley at Cambridge.

[12] *Gardener's Magazine,* XVIII (1842), 290-291, 296-301.

At Honolulu, Douglas's grave went unmarked (except for a brick covering which soon disintegrated) until 1855, when an English visitor, Rev. Julius L. Brenchley, took matters in hand. Unable to obtain a suitable tombstone there, he got a marble one in San Francisco on his way home, had it inscribed in Latin, and sent it across—all at his own expense. As the whereabouts of the grave were uncertain it was erected on an outside wall of the Kawaiahao church.[13]

In 1929 the Royal Horticultural Society learned that the stone's inscription was becoming illegible, so had the stone removed to the vestibule of the church and had a bronze tablet placed below it containing a copy of the inscription with the following translation:

> Here lies Master David Douglas, born
> in Scotland A. D. 1799. An indefati-
> gable traveller, he was sent out by
> the Royal Horticultural Society of
> London, and gave his life for Science
> in the wilds of Hawaii, July 12, 1834.
> "E'en here the tear of pity springs
> And hearts are touched by human things."
> —Virgil.[14]

The spot where Douglas was killed (named by the natives *Kaluakauka*, "the Doctor's Pit"), has been marked by a stone cairn and a grove of young Douglas firs, both of which were dedicated to his memory by the Hilo Burns Club on July 12, 1934, just one hundred years after his death.

Honored as Douglas is by these memorials of stone and metal, with most people his memory will be kept alive by the plants, shrubs, and trees that he introduced. Gardens and grounds throughout the world abound with them. Hardly a garden exists that does not have the clarkia, mariposa lily, California

[13] John Goldie, "In Memory of David Douglas," pp. 89-94.

[14] *Ibid.; Journal of the Royal Horticultural Society*, vol. LVI (1931), p. lxxx and fig. 82-84.

poppy, or some of the lupines, phlox, penstemons, mimulus, or others of his beautiful flowers.

Towering over all is the great tree, the Douglas fir, in connection with which he is chiefly remembered. In addition to seeds, it appears that he sent home samples of the wood, for in 1827 John Lindley of the Horticultural Society received two planks each about twenty feet long to test for durability.[15] Strong, tough, straight-grained, durable, and hardening with age, it has been proven unsurpassed in the qualities which render lumber most valuable. It has a great variety of uses, but because of its strength and size it is particularly valuable for heavy construction, and has become the world's greatest structural timber. Growing over a wide range of territory, thriving under diverse conditions, and attaining large dimensions—trees 250 feet high and from six to nine feet in diameter are frequently encountered —it is the most important tree in the American lumber trade.

Outstanding also as an ornament, this "King of the Conifers" is a favorite with garden lovers and tree planters. It has been introduced into gardens and estates wherever climatic conditions permit its growth; but nowhere outside its original habitat has it grown to greater perfection than in Perthshire, where it is deservedly popular. The Earl of Mansfield, Douglas's old employer, was one of the earliest planters; and today, in the Scone Palace grounds near his birthplace, may be seen a sturdy Douglas fir raised from the first seeds brought home by him in 1827.

It is also distinguished for the large number of names by which it is or has been called—more than those of any other American tree. Oregon pine, red pine, Puget Sound pine, Oregon spruce, red spruce, Douglas spruce, red fir, yellow fir, Oregon fir, and spruce fir are some of the common names. Scientists have fought over the name more than laymen. The name given by Lambert, *Pinus Douglasii,* came to be *Pseudotsuga Douglasii* Carriere, the name in general use in Great

[15] R. C. Mayne, *Four Years in British Columbia,* p. 409.

Britain. According to the code of the International Botanical Conference of 1905 and subsequent Conferences, however, the name is *Pseudotsuga taxifolia* (Lambert) Britton, which is the name officially recognized in Canada and becoming generally adopted in the United States.[16]

It has been said that the best way to get at a man's character is to study his love letters. Written off guard and in an intimate and confidential vein, they reveal his inner character. From that view David Douglas is a disappointment. He wrote no love letters, had no love affairs. There was no woman in his life except his mother. He was occupied with other and, to him, more important things. His attention was directed toward the beauty of the lower orders of nature which graced forest and prairie, swamp and ravine, river bank and mountainside, toward the grandeur of streams and waterfalls, turquoise lakes, mountain peaks, fiery volcanoes, and the starry firmament. Consequently those were the things he wrote about, the objects of his affection.

Nevertheless, the letters he did write are frank and open. His journals, too, are intimately written and self-revealing. Sometimes in them he talks to himself. From these and the other sources of information regarding him it is plain that the dominant trait of his character was devotion to duty: duty to the Horticultural Society, his employer; duty to Hooker, his teacher and friend; duty to natural history, his only mistress.

A votary of science, Gairdner called him. From the long, hard-working days of his youth as a garden boy, his whole life was given to the pursuit of science and to that alone. It was his love and his satisfaction. As James K. Townsend, the American

[16] John Davidson, *Conifers, Junipers, and Yew: Gymnosperms of British Columbia* (London, 1927), pp. 39-40; Department of the Interior, Canada, *Native Trees of Canada*, Forest Service Bulletin No. 61 (revised edition, Ottawa, 1933), pp. 6, 61; *Journal of the Royal Horticultural Society*, XIV (1892), v, 12-13, 245. In the case of many other plants scientific names honoring Douglas have been retained. At least seventeen of the Pacific Northwest plants still bear his name.

naturalist who followed him on the Columbia, has said of the resources of the naturalist, "the wild forest, the deep glen, and the rugged mountain top, possess charms for him which he would not exchange for gilded palaces; and . . . to acquaint himself with nature he gladly escapes from the restraints of civilization, and buries himself from the world which cannot appreciate his enjoyment." [17]

Concomitant with his zeal were his bravery and perseverance. He faced dangers, overcame difficulties, endured hardships, and made sacrifices such as have been the experience of few men. How many would not have given up in despair upon losing in a river-crossing "a multitude of treasures botanical and zoological . . . the labour of fifty-four days of fatigue and anxiety!" [18] How many could have carried an eagle across the continent, and another halfway, and lost both at the journey's end, without bitter complaint!

Douglas got along well with the Indians. Cool-headed and tactful, he knew how to treat them. He could be firm or be kind, as the occasion required. On some journeys they were his only companions. He ate their food, spoke their language.

He was eminently successful in his work. "There is scarcely a spot deserving the name of a garden, either in Europe or the United States," says an English admirer, "in which some of the discoveries of Douglas do not form the chief attraction; . . . to no single individual is modern horticulture more indebted than to David Douglas." [19] Some of the flowers became better known in European gardens than in their native country. The pretty farewell-to-spring (*Godetia amoena*) which he introduced into England was returned to America in many improved and charming cultivated varieties. The glowing yellow *Eriophyllum lanatum* (*E. caespitosum* Dougl.), which has been widely cultivated

[17] Townsend, *Narrative*, p. 214.

[18] Douglas, in *Transactions*, Linnean Society, XVI (1833), 145.

[19] An unnamed writer quoted by Neville Cooper in *Gardener's Chronicle* (Sept. 25, 1926).

in Europe since he introduced it, is only now (under the name of Oregon sunshine) becoming valued by American gardeners.[20]

As the first botanist to systematically comb the Pacific Northwest and California, Douglas made their flora known to the world and prepared the way for others. Following him came Townsend and Nuttall on the Columbia, and Karl T. Hartweg and William Lobb in California, while later came John Jeffrey, collecting from the Fraser River to California.

His pioneering was, of course, not confined to botany. As the first scientist of any kind to penetrate inland in the Northwest, he did not neglect opportunities in other branches of science. He realized the danger of confining himself too closely to one thing:

To console myself for the want of friends of a kindred feeling in this distant land, for an exchange of sympathy or advice, I vary my amusements; by day it is a barren place that does not afford me a blade of grass; a bird, or a rock, before unnoticed, from which I derive inexpressible delight, while during the stillness of a cloudless night their localities are determined, altitudes measured, the climate they breathe analyzed. Thank God my heart feels gladness in these operations; without such to pass away an hour, my time would be blank. I willingly admit, however, there is one way, and one only way in which a man's powers may be cramped by the pursuit of natural truth, and that is by too ardent devotion to it. In the pursuit of any subject, however lofty, a man may become narrow-minded, and in a condition little better than moral servitude, but by embracing different subjects, we need not fear on this head. Every department of science offers its spoils for our decoration. We can be carried into regions where we contemplate the most glorious workmanship of nature, and where the dullest imagination becomes excited. We can travel through distant lands and become acquainted with the complexions and the feelings and the characters of mankind, under every form of life; and in doing this, if we be not most indocile pupils we must learn many lessons of kindness, and freedom

[20] Haskin, "David Douglas."

of thought along with appropriate knowledge of our immediate vocation.[21]

In zoology his interest was chiefly with birds. Many of his specimens decayed owing to the long voyage or other causes, but several reached the museums at the University of Edinburgh, the Andersonian Institution at Glasgow, and the Zoological Society of London. Many of these are mentioned by early ornithologists. Others were sent to James Wilson for his *Illustrations of Zoology*.

On his last American expedition much of Douglas's attention was given to the observations for which Captain Sabine had instructed him. He determined the latitude and longitude of many places on the Pacific Slope and in the Sandwich Islands, and the elevations of various mountains. Sabine was particularly pleased with the magnetic observations. They gave him great assistance in substantiating the theory that there are two magnetic poles in the Northern Hemisphere, and that the one in North America has a sensibly greater intensity than that in Siberia.[22]

Douglas also made history as a mountain climber. He was the first white man known to climb one of the Canadian Rocky Mountains; first to climb any of the Cascade Mountains; first to climb the Blue Mountains of Oregon; first to ascend both the great Hawaiian peaks; and first to attempt Mount Hood.

He did his work with thoroughness and care, be it exploring, collecting, observing, or recording. He had a good command of English; his journals and letters are vigorously and fluently written, and his professional papers are lucid. The records of his observations showed great care in both taking and recording.

Although killed while still a young man, Douglas had seen much of life. He had varied his vocation, traveled widely, had strange experiences, and seen wonderful sights; he had dealt

[21] Letter, March 17, 1833, Douglas to Barnston, in Barnston, "Abridged Sketch," p. 277.

[22] Edward Sabine, "Report," pp. 27-32, 63-85.

with men of all ranks and of various races, used diverse means of travel and several languages, and lived under makeshift conditions and on peculiar foods. Certainly, his career did not lack scope and variety. If life be measured by achievement rather than by time, the life-cup of David Douglas was filled. Into the fraction of an ordinary life-period that fate allowed him he had crowded activities which might have occupied a full lifetime. He met his end young in age, but mature in achievement.

APPENDICES

APPENDIX A

ACCIDENTAL DEATH OR MURDER?

Since the question of how Douglas met his death has caused much discussion, further consideration is given here.

THE DOUBTS and suspicions that were in the minds of Douglas's friends after his death were not dispelled by the opinion of the doctors. Ned Gurney had a poor reputation. His explanations were not generally accepted, and many persons shared Hall's view that he had made away with Douglas for his money. After robbing him he might have pushed him into the pit, in which case the medical verdict would be reconcilable. It was alleged that Gurney bought passage from the Islands with the "blood-money" soon afterwards.[1]

Others had similar suspicions against John. He disappeared completely and was never heard of again. Whether he lost his way after Douglas left him and perished on the mountain, or whether he ran away, was never found out.[2]

Fifteen months after Douglas's death, Dr. Meredith Gairdner, then in Honolulu for his health, found "the minds of the residents and particularly of the British Consul . . . by no means satisfied that he came by his end by mere accident; they say that it is impossible for a man in daylight to fall into one of these pits in which a bullock is trapped, for the animal breaks down the whole covering when he falls in leaving a large and visible aperture. D. had latterly become very irritable and had frequent quarrels with the natives. It is curious that the chronometer he carried in his breastpocket secured in a brass case was smashed to atoms, but the case uninjured. While the other in his fob was still going when the body was found altho'

[1] Another report is that he remained until 1849, when he joined the California gold-rush. A third report says he stayed in Hawaii until his death (Burns, "The David Douglas Mystery").

[2] *Ke Kumu Hawaii* (newspaper) Honolulu, Nov. 26, 1834, pp. 13-15; Wilkes, *Narrative*, IV, 204.

the fob and all that side of his dress was torn to shreds. The few dollars he had were not taken away."[3]

More than six years after the tragedy the scene was visited by Dr. Charles Pickering and J. D. Brackenridge, of the United States Exploring Expedition commanded by Lt. Charles Wilkes. They found the bullock pit and made a cursory investigation but expressed no conclusion, remarking, however, that "few deaths could be more awful than that which he is supposed to have suffered." They pointed out how easy it was for a person to fall into a cattle pit and told of a native "who, having prepared a pit, succeeded in entrapping a large bull, but became so excited at his success, that he slipped and fell in himself; however, being armed with a knife, he succeeded in killing the animal. When discovered both were dead."[4]

Rev. Titus Coan, well-known missionary, did not arrive until the year after Douglas's death. However, he heard enough regarding the mystery to lead him to write as follows: "The dead body of the distinguished Scotch naturalist Douglass, was found under painfully suspicious circumstances, that led many to believe that he had been murdered for his money. A mystery hangs over the event which we are unable to explain."[5]

After the lapse of more than forty years it remained for a Hilo newspaper to come forward with an avowed solution of the mystery. In an article headed "DEATH OF PROF. DOUGLAS. A Bit of History, a Doubt Cleared up as to His Death," the *Hilo Tribune* of August 22, 1896, proclaimed that "as murder will out, so now after many days, almost forgotten in the vanishing years, the tragical circumstances of his death have become invested with new interest, and for the first time there is published to the world a story of foul, black and cowardly murder." Then follows the report of a conversation with Bolabola, "a noted hunter and trailer, over 70 years of age, familiar with every inch of ground around the Mauna Kea slopes," who said that at the time of Douglas's death he was about ten years old and lived with his parents at Lahohinu, near Ned Gurney's. After giving a garbled version of the tragedy, which conflicts with the missionaries' report and other known facts, he concluded: "The haole [foreigner] was murdered, we all felt so at the time, but we were afraid to say so and only whispered it among ourselves. And when

[3] Letter, Nov. 19, 1835, Gairdner to W. J. Hooker, Kew Gardens.
[4] Wilkes, *Narrative*, IV, 204-205.
[5] Titus Coan, *Life in Hawaii* (New York, 1882), p. 225.

my father and old Kaline (another noted Hawaiian) died, they both repeated the story to me."[6]

Although this disappointing article added nothing, explained nothing, nor suggested who the murderer was, nevertheless for some it confirmed belief in the murder theory.[7]

In many ways the accident theory seems more consistent with Douglas's characteristics and behavior: his impatiently declining to wait for the natives who were going his way and confidently proceeding alone; his inquisitively peering into the bullock pit; and, perhaps owing to his poor eyesight, his getting too close to the edge.

Billy's behavior also must not be overlooked. It is hard to reconcile with murder. If Douglas had been attacked and thrown into the pit the faithful terrier probably would have been drawn into the struggle, and shown signs of it—perhaps suffered the same cruel end as his master.

[6] The article, unsigned, follows the editorials. The editor of the newspaper then was Gardner K. Wilder, who died in 1904. The newspaper is at the Hawaiian Historical Society Library, Honolulu.

[7] Hitchcock, *Hawaii and Its Volcanoes*, pp. 81-82; L. W. Bryan, "Kaluakauka," p. 30. The footnote in Coan's book referred to in the latter was probably written by his son, Titus Munson Coan.

APPENDIX B

PLANTS

INTRODUCED BY DAVID DOUGLAS

Condensed from the Royal Horticultural Society's list in David Douglas, *Journal,* London, 1914, pp. 326-336. Species there undetermined have been omitted, as have also species there determined as being the same as others in the list. Some names have been changed since the list was made, and in a few cases the names now accepted are given here in parentheses; but no attempt has been made to give them all, for botanical nomenclature is a complex subject. Common names have been added with the kind assistance of Mr. J. W. Eastham, plant pathologist, British Columbia Department of Agriculture, Vancouver; but here, too, there is much variation. The marginal letters, B for Brazil, C for California, indicate the places where the plants were found by Douglas. Species not lettered came almost entirely from the Pacific Northwest. It must be remembered that "California," as used by Douglas, included what is now southern Oregon.

	Latin Name	Common Name
	Abies amabilis	Lovely fir
	grandis	White fir
	nobilis	Noble fir
	Abronia mellifera	Honey-scented sand verbena
	Acer circinnatum	Vine maple
	macrophyllum	Broad-leaved maple
	Amelanchier alnifolia	Saskatoon or Juneberry
C	*Amsinckia lycopsoides*	Fiddle-neck
	Anemone multifida (A. globosa)	Globose anemone
C	*Antirrhinum glandulosum*	Sticky snapdragon
	Arbutus Menziesii	Madrona
	Arctostaphylos tomentosa	Manzanita
C	*Aster canescens var. viscosus*	Hoary aster
	Astragalus caryocarpus	A Milk vetch, rattle weed, or loco weed
	Audibertia incana	Gray-ball sage

	Latin Name	Common Name
	Berberis Aquifolium (Mahonia Aquifolium)	Tall Oregon grape
	nervosa (Mahonia nervosa)	Low Oregon grape
B	*Brassavola nodosa*	A tropical dwarf orchid
	Brodiaea congesta	Wild hyacinth (five species)
	grandiflora	
	ixioides	
C	*lactea*	
C	*laxa*	
	Calandrinia Menziesii	Redmaids
C	*Calochortus albus*	Mariposa lily or tulip (six species)
C	*luteus*	
	macrocarpus	
C	*pulchellus*	
C	*splendens*	
C	*venustus*	
	Camassia esculenta	Camas
	Castilleja parviflora	Small-flowered paint-brush
	Chelone nemorosa	Turtlehead
	Clarkia elegans	Clarkia (three species)
	pulchella	
C	*rhomboidea*	
	Clematis virginiana	Virginia clematis
C	*Collinsia bicolor*	Chinese houses
	grandiflora	Blue-eyed Mary
	parviflora	
	Collomia gracilis	Collomia
	grandiflora	
	linearis	
	Coreopsis Atkinsoniana	Atkinson's Coreopsis or tickseed
	Cornus stolonifera	Red-osier dogwood
	Crataegus Douglasii	Hawthorn
	Delphinium Menziesii	Menzies' larkspur or low larkspur
C	*Dendromecon rigidum*	Bush poppy
	Douglasia nivalis	Mountain pink
	Downingia elegans	Downingia
C	*pulchella*	
	Epilobium minutum	Willow herb
C	*Erigeron speciosus*	Showy fleabane

	Latin Name	Common Name
	Eriogonum compositum nudum	Sulphur flower
	Eriophyllum caespitosum (E. lanatum)	Woolly sunflower
	Erythronium grandiflorum	Dog-tooth violet or glacier lily
C	*Eschscholzia californica*	California poppy (four forms)
C	*californica var. caespitosa*	
C	*californica var. Douglasii*	
C	*californica var. hypecoides*	
	Fritillaria pudica	Mission bell
	Gaillardia aristata	Gaillardia
C	*Garrya elliptica*	Quinine or tassel bush
	Gaultheria shallon	Salal
	Geranium carolinianum	Carolina geranium
B	*Gesneria Douglasii*	A tropical American herb
C	*Gilia archilleaefolia*	Gilia (eleven species)
	aggregata	Scarlet gilia
C	*androsacea*	Feather gilia
	capitata	Blue gilia
	coronopifolia	
C	*densiflora*	
C	*liniflora var. pharnaceoides*	
C	*multicaulis*	
	pungens	
C	*tenuiflora*	
C	*tricolor*	Birdseye gilia
B	*Gomesa planifolia*	An epiphytic orchid
	Helianthus annus	Sunflower
	Heuchera cylindrica	Alum root
	micrantha	Small-flowered alum root
	Hosackia bicolor	Bicolored bird-foot clover
C	*crassifolia*	Pink bird-foot clover
	Iris tenax	Tough-leaved iris
C	*Lasthenia glabrata*	Lasthenia
	Lathyrus maritimus	Beach pea
C	*Layia calliglossa*	Tidy tips
	Lespedeza capitata	Bush clover
C	*Limnanthes Douglasii*	Meadow foam

	Latin Name	Common Name
	Linum perenne	Perennial blue flax
	Lonicera ciliosa	Orange honeysuckle
	hirsuta	Hairy honeysuckle
	hispidula	Purple honeysuckle
	Lophanthus urticifolius	Giant hyssop
	Lupinus aridus	Lupine (twenty-one species)
C	*Chamissonis*	
C	*densiflorus*	
	flexuosus	
C	*hirsutissimus*	
	laxiflorus	
	lepidus	
C	*leptophyllus*	
	leucophyllus	
	littoralis	
	micranthus	
	micranthus var. bicolor	
C	*nanus*	
C	*polyphyllus*	
	polyphyllus var. albiflorus	
C	*polyphyllus var. grandifolius*	
C	*rivularis*	
C	*rivularis var. latifolius*	
	ornatus	
	Sabinii	
	sulphureus	
	Madia elegans	Showy madia
	Malvastrum Munroanum	False mallow
C	*Meconopsis heterophylla*	Flaming or wind poppy
C	*Mentzelia Lindleyi*	Blazing-star
B	*Mikania scandens*	Climbing hempweed
	Mimulus cardinalis	Mimulus or monkey-flower (five species)
	floribundus	
C	*Lewisii*	Pink mimulus
	luteus	Golden mimulus
	moschatus	Musk flower
C	*Nemophila aurita*	Climbing nemophila

	Latin Name	Common Name
C	*Nemophila insignis*	Baby-blue-eyes
	Nicotiana quadrivalvis var. multivalvis	Native tobacco
	Oenothera albicaulis	Evening primrose (eleven species)
	amoena (Godetia amoena)	
	biennis var. muricata	
C	*decumbens*	
C	*densiflora*	
	dentata	
C	*lepida*	
	quadrivulnera	
	speciosa	
	triloba	
C	*viminea*	
B	*Oncidium pubes*	A tropical epiphytic orchid
	Paeonia Brownii	American or wild peony
	Pedicularis canadensis	Lousewort
	Penstemon acuminatus	Beard-tongue or penstemon (eighteen species)
	attenuatus	
C	*breviflorus*	
C	*centranthifolius*	
	confertus	
	deustus	
	diffusus	
	glaber	
	glandulosus	
	gracilis	
C	*heterophyllus*	
	Menziesii var. Douglasii	
	Menziesii var. Scouleri	
	ovatus	
	pruinosus	
	Richardsonii	
	triphyllus	
	venustus	
C	*Phacelia divaricata*	Wild heliotrope or phacelia (four species)
	Menziesii	

	Latin Name	Common Name
C	*Phacelia tanacetifolia*	
C	*viscida*	
	Phlox linearifolia	Phlox
	Picea sitchensis	Sitka spruce
C	*Pinus Coulteri*	Coulter or big-cone pine
C	*insignis (P. radiata)*	Monterey pine
	Lambertiana	Sugar pine
	monticola	Western white pine
	ponderosa	Western yellow pine
C	*Sabiniana*	Gray or Digger pine
C	*Platystemon californicus*	Creamcups
C	*Platystigma lineare*	Platystigma
	Pogonia pendula	A small terrestrial orchid
	Potentilla arguta (P. valida)	Cinquefoil or five-finger (six species)
C	*congesta*	
	effusa	
	glandulosa	
	pensylvanica	
	pensylvanica var. strigosa	
	Prunus pumila	Sand cherry
	Pseudotsuga Douglasii	Douglas fir
C	*Psoralea macrostachya*	Leather-root
C	*orbicularis*	
	Purshia tridentata	Antelope bush
	Pyrus rivularis (Malus rivularis)	Pacific coast crab-apple
	Ribes aureum var. tenuiflorum	Golden currant
	cereum	Squaw currant
	divaricatum	Straggly gooseberry
	divaricatum var. irriguum	Rock or black gooseberry
	gracile	Pasture gooseberry
	Hudsonianum	Western black currant
	lacustre	Swamp currant
	Menziesii	Fog or canyon gooseberry
	Ribes sanguineum	Red-flowered currant
C	*sanguineum var. glutinosum*	Winter currant

	Latin Name	Common Name
C	*Ribes sanguineum var. malvaceum*	Chaparral currant
	setosum	Red-shoot gooseberry
C	*speciosum*	Fuchsia-flowered gooseberry
	viscosissimum	Sticky currant
	Rubus leucodermis	Black raspberry
	nutkanus (R. parviflorus)	Thimbleberry
	spectabilis	Salmonberry
	Sidalcea malvaeflora	Marsh hollyhock
B	*Sinningia Helleri*	A gloxinia
	Spergula arvensis	Common spurrey
	Spiraea Aruncus (Aruncus sylvester)	Goat's beard
	discolor var. ariaefolia	Ocean spray
	Symphoricarpos racemosus	Snowberry or waxberry
	Tanacetum huronense	Huron tansy
C	*Trifolium fucatum*	Puff clover or sour clover
	Vaccinium ovatum	Evergreen huckleberry
C	*Xerophyllum asphodeloides (X. tenax)*	Bear grass

In all, 215 plants.

BIBLIOGRAPHY

THE CHIEF AUTHORITIES for the story of Douglas's life are the published journals of his travels and the published extracts from his letters, principally those to Doctor (afterwards Sir) William Jackson Hooker, professor of botany at Glasgow. The journals begin with that kept on his visit to the United States and Upper Canada in 1823. Of his next journey, the first expedition to the Columbia (1824-1827), three different accounts have been published, all from material written by Douglas and in the possession of the Horticultural Society of London, or its successor, the Royal Horticultural Society, London. The longest of these is a composite of two manuscripts, one of which is presumably the original journal written in the course of his journey, while the other is probably made up of transcripts of parts of the original which he sent from time to time to his employer, the Horticultural Society of London, to serve as reports of his work. The other two accounts are: (1) extracts from a revision of the original journal which was probably prepared for the book John Murray proposed to publish, but for some reason did not publish; and (2) a very condensed revision, made possibly from the former. Of Douglas's last expedition to the Western Hemisphere (1829-1834) only a small portion of a journal has ever been found. In the course of this expedition he severed his connection with the Horticultural Society, with the result that he no longer sent a transcript to London. The journal up to June, 1833, was lost when he was wrecked in the Fraser River. The portion that was found of the remainder describes his ascents of the volcanoes on the island of Hawaii during January, 1834.

The first important publication of Douglas material was Hooker's "Brief Memoir of the Life of Mr. David Douglas, with Extracts from His Letters," published in the *Companion to the Botanical Magazine*, volume II (London, 1836; reprinted in the *Hawaiian Spectator*, volume II, Honolulu, 1839, and in the *Oregon Historical Quarterly*, volumes V and VI, Portland, Oregon, 1904 and 1905). This begins with a biographical sketch of Douglas up to the time of his first Columbian expedition in 1824. Then come extensive selections from what is believed to be the revision of his journal which Douglas prepared for Murray, entitled "A Sketch of a Journey to the North-Western Parts of the Continent of North America during the Years 1824, 5, 6, and 7." This is followed by a brief account by Hooker of Douglas's next two years in London. Information regarding the final expedition during 1829-1834 is given in extracts from Douglas's letters to Hooker, and in the found portion of his journal (which was lent to Hooker by Douglas's brother, John). The circumstances of Douglas's death are told in a letter written by two American missionaries in Hawaii to Richard Charlton, British Consul at Honolulu.

The second important publication was that directed by the Royal Horticultural Society, entitled *Journal Kept by David Douglas during His Travels in North America, 1823-1827*, published at London in 1914. This begins with Douglas's journal of his visit to eastern America in 1823, followed by his account of the American oaks found on this visit. It continues with two accounts of the first expedition to the Columbia. The first of these is the very condensed journal revision already mentioned, which has the same title and opening paragraph as the version in Hooker's "Memoir," but is much shorter. The second is the long, composite account also mentioned above. It is the most complete of the three accounts, although each of them contains material not in the others, and there are discrepancies between them. The Royal Horticultural Society's book makes no mention of Douglas's two years in England between expeditions, and for the last expedition (1829-1834) it has only the found portion of his journal and the letter of the missionaries, both taken from Hooker's "Memoir." It also contains a list entitled "Plants introduced by David Douglas during the years 1826-34," and a composite of two accounts by Douglas entitled "Some American Pines."

General authorities for particular chapters are as follows:

CHAPTER 1—John Douglas, "Biographical Notes of David Douglas," MS., Kew Gardens; Fish (initials unknown), letter in *Cottage Gardener*, London, VI (1851), 263; W. J. Hooker, "A Brief Memoir of the Life of Mr. David Douglas," *Companion to the Botanical Magazine*, London, II (1836), 79-82 (reprinted in *Hawaiian Spectator*, Honolulu, vol. II, 1839, pp. 2-8, and *Oregon Historical Quarterly*, Portland, Ore., vol. V, 1904, pp. 223-229. Page numbers given in subsequent references to Hooker's "Memoir" are those for the *Oregon Historical Quarterly*); J. D. Urquhart, *Historic Sketches of Scone* (Perth, 1883).

CHAPTER 3—David Douglas, *Journal Kept by David Douglas during*

His Travels in North America, 1823-1827 (London, 1914), pp. 1-30; John T. Faris, *Old Gardens in and about Philadelphia and Those Who Made Them* (Indianapolis, 1932); Edwin C. Guillet, *Early Life in Upper Canada* (Toronto, 1933); Ulysses Prentiss Hedrick, *A History of Agriculture in the State of New York* (Albany, 1933); U. P. Hedrick, *The Peaches of New York* (Albany, 1917); U. P. Hedrick, *The Plums of New York* (Albany, 1911); John Howison, *Sketches of Upper Canada* (Edinburgh, 1821); Max Meisel, *Bibliography of American Natural History* (Brooklyn, N. Y., 1926), vol. II.

CHAPTER 4—Douglas, *Journal*, pp. 50-56, 77-103; David Douglas, "Sketch of a Journey to the Northwestern Part of the Continent of North America during the Years 1824-25-26-27," in W. J. Hooker, "Memoir," *OHQ*, V, 230-243.

CHAPTER 5—Douglas, *Journal*, pp. 56-60, 103-146; Douglas, "Sketch," *OHQ*, V, 243-252.

CHAPTER 6—Douglas, *Journal*, pp. 60-62, 146-156; Douglas, "Sketch," *OHQ*, V, 252-260.

CHAPTER 7—Douglas, *Journal*, pp. 62-63, 157-189; Douglas, "Sketch," *OHQ*, V, 260-271, 325-349.

CHAPTER 8—Douglas, *Journal*, pp. 64-66, 188-212; Douglas, "Sketch," *OHQ*, V, 349-369, VI, 76.

CHAPTER 9—Douglas, *Journal*, pp. 66-69, 212-241; Douglas, "Sketch," *OHQ*, VI, 76-97.

CHAPTER 10—Douglas, *Journal*, pp. 69-76, 241-293; Douglas, "Sketch," *OHQ*, VI, 206-218; Edward Ermatinger, "York Factory Express Journal," *Transactions*, Royal Society of Canada, Ottawa, 3rd ser., Vol. VI (1912), sec. II, pp. 70-93.

CHAPTER 14—Letters of David Douglas to W. J. Hooker, in W. J. Hooker, "Memoir," *OHQ*, VI, 293-302; Edward Sabine, "Observations taken on the Western Coast of North America by the late Mr. David Douglas, with a Report on his Papers, 1837," MS., Royal Society, London; letter, Nov. 11, 1833, David Douglas to W. E. P. Hartnell, in *Madrono*, Journal of the California Botanical Society, Berkeley, Cal., II (1933), 98-100.

CHAPTER 15—Letters of Douglas to W. J. Hooker, in W. J. Hooker, "Memoir," *OHQ*, VI, 297-305.

CHAPTER 16—Letters of Douglas to W. J. Hooker, in W. J. Hooker, "Memoir," *OHQ*, VI, 223, 225, 296, 302-309; Edward Sabine, "Observations."

CHAPTER 18—Douglas, letters to W. J. Hooker, and "Voyage," in W. J. Hooker, "Memoir," *OHQ*, VI, 307-309, 417-441 (Note: The section of Hooker's "Memoir" appearing on pages 419-441 of *OHQ*, VI, consists of selections from the only fragment found of Douglas's journal of his last voyage. It appeared in the original printing of Hooker's "Memoir" under the heading "Mr. Douglas' Voyage from the Columbia to the Sandwich

Islands, and the Ascent of Mouna Roa." This section of the "Memoir" is referred to in the footnotes as: Douglas, "Voyage."); letter, May 3, 1834, David Douglas to Edward Sabine, *Journal*, Royal Geographic Society, London, IV (1834), 333-344; letter by David Douglas printed in *Sandwich Island Gazette*, Honolulu (July 30, 1836).

CHAPTER 19—W. J. Hooker, "Memoir," *OHQ*, VI, 441-449; Douglas, *Journal*, pp. 318-323; Frederick Debell Bennett, *Narrative of a Whaling Voyage Round the Globe from the Year 1833 to 1836* (London, 1840), I, 282-284; Mrs. D. B. Lyman, Journal, MS., Hawaiian Mission Children's Society Library, Honolulu; Levi Chamberlain, Journal, MS., Hawaiian Mission Children's Society Library, Honolulu.

BOOKS AND MAGAZINE ARTICLES

Acton, Frances, *see* Thomas Andrew Knight.

Baillie-Grohman, W. A., *Fifteen Years' Sport and Life in the Hunting Grounds of Western America and British Columbia* (London, 1900).

Ball, John, "Across the Continent Seventy Years Ago," extracts from the journal and letters of John Ball compiled by his daughter, Kate N. B. Powers, *Oregon Historical Quarterly*, Portland, Ore., III (1902), 82-106.

———, *Autobiography of John Ball*, compiled by his daughters, Kate Ball Powers, Flora Ball Hopkins, and Lucy Ball (Grand Rapids, Mich., 1925).

Balfour, F. R. S., "David Douglas," *Journal of the Royal Horticultural Society*, London, LXVII (1942), 121-128, 153-162.

Bancroft, Hubert Howe, *History of British Columbia* (San Francisco, 1890).

———, *History of California*, 7 vol. (San Francisco, 1886-1890).

———, *History of the Northwest Coast*, 2 vol. (San Francisco, 1884).

———, *History of Oregon*, 2 vol. (San Francisco, 1886-1888).

———, *History of Washington, Idaho and Montana* (San Francisco, 1890).

Barnston, George, "Abridged Sketch of the life of Mr. David Douglas, Botanist, with a few details of his travels and discoveries," *Canadian Naturalist and Geologist*, Montreal, V (1860), 120-132, 200-208, 267-278, 329-349.

Beale, Thomas, *The Natural History of the Sperm Whale* (London, 1839).

Beechey, F. W., *Narrative of a Voyage to the Pacific and Beering's Strait . . . in the Years 1825, 26, 27, 28*, 2 vol. (London, 1831).

Belcher, Edward, *Narrative of a Voyage Round the World . . . during the Years 1836-1842*, 2 vol. (London, 1843).

Bennett, Frederick Debell, *Narrative of a Whaling Voyage Round the Globe from the Year 1833 to 1836*, 2 vol. (London, 1840).

Bourinot, John, *see* Nicholas Garry.

Bower, F. O., "Sir William Hooker," in *Makers of British Botany,* F. W. Oliver, ed. (Cambridge, England, 1913).

British Columbia, Department of Lands, Lands and Survey Branch, *Annual Report* for 1936 (Victoria, B. C., 1937).

Bryan, L. W., "Kaluakauka," *Paradise of the Pacific,* Honolulu, December, 1934, pp. 28-30.

Bryan, William Alanson, *Natural History of Hawaii* (Honolulu, 1915).

Burns, Eugene, "The David Douglas Mystery," *Honolulu Star-Bulletin,* Honolulu, April 23, 1938.

Burpee, Lawrence J., *On the Old Athabaska Trail* (London, 1927).

Canada, Department of the Interior, *Native Trees of Canada,* Forest Service Bulletin No. 61 (revised edition, Ottawa, 1933).

Canada, Geological and Natural History Survey, Annual Reports (Montreal).

Canadian Northwest, see E. H. Oliver.

Chittenden, F. J., "The Lost Scent of Musk," *Gardener's Chronicle,* London, LXXXVIII (1930), 457.

Clark, William, *see* Meriwether Lewis.

Coan, Titus, *Life in Hawaii* (New York, 1882).

Coleman, Arthur P., *The Canadian Rockies, New and Old Trails* (London, 1911).

Companion to the Glasgow Botanic Garden (Glasgow, c. 1818).

Cox, Ross, *Adventures on the Columbia River,* 2 vol. (London, 1831).

Dana, James D., *Characteristics of Volcanoes* (New York, 1890).

———, "Geology," vol. X of *United States Exploring Expedition . . . under the Command of Charles Wilkes, U.S.N.* (Philadelphia, 1849).

Davidson, John, *Conifers, Junipers and Yew: Gymnosperms of British Columbia* (London, 1927).

Davis, William Heath, *Seventy-five Years in California* (San Francisco, 1929).

Dawson, G. M., "The Mineral Wealth of British Columbia," in *Annual Report* for 1887-88, Geological and Natural History Survey of Canada (Montreal, 1889).

De Smet, Pierre Jean, *see* Smet.

Douglas, David, *Journal Kept by David Douglas during His Travels in North America, 1823-1827* (London, 1914).

———, "A Sketch of a Journey to the North-Western Parts of the Continent of North America, during the Years 1824, 5, 6, and 7," selections from a revision by Douglas of his own journal, printed as part of W. J. Hooker's "Brief Memoir of the Life of Mr. David Douglas, with Extracts from His Letters," *Companion to the Botanical Magazine,* London, II (1836), 79-182, reprinted in the *Hawaiian Spectator,* Honolulu, II (1839) and in the *Oregon Historical Quarterly,* Portland, Ore., V and VI (1904 and 1905). Volume and page references in the footnotes refer to the *Oregon Historical Quarterly.*

———, "Mr. Douglas' Voyage from the Columbia to the Sandwich Islands, and the Ascent of Mouna Roa," selections from the only part that has been found of Douglas's journal of his last voyage to the Western Hemisphere, printed in W. J. Hooker's "Memoir" (*see* Douglas, "Sketch," above).

Drummond, Thomas, "Sketch of a Journey to the Rocky Mountains and to the Columbia River in North America," in W. J. Hooker's *Botanical Miscellany*, London, I (1830), 178-219.

Drury, Clifford M., "Oregon Indians in the Red River School," *Pacific Historical Review*, Glendale, Cal., VII (1938), 50-60.

Eastwood, Alice, "Early Botanical Explorers on the Pacific Coast and the Trees They Found There," *California Historical Society Quarterly*, San Francisco, XVIII (1939), 341.

Elliott, T. C., *see* John Work.

Engelmann, George, "Notes on Western Conifers," *Botanical Gazette*, Indianapolis, VII (1882), 4-5.

———, "A Synopsis of American Firs," *Transactions of the Academy of Sciences of St. Louis*, III (1868-1877), 593-602.

Ermatinger, C. O., *see* Edward Ermatinger.

Ermatinger, Edward, "Edward Ermatinger's York Factory Express Journal, Being a Record of Journeys Made Between Fort Vancouver and Hudson Bay in the Years 1827-1828," C. O. Ermatinger and James White, editors, *Proceedings and Transactions of the Royal Society of Canada*, Ottawa, 3rd ser., vol. VI (1912), sec. II, pp. 67-132.

Faris, John T., *Old Gardens in and about Philadelphia and Those Who Made Them* (Indianapolis, Ind.: Bobbs-Merrill Co., 1932).

Fish (initials unknown), letter in *Cottage Gardener*, London, VI (1851), 263.

Franklin, John, *Narrative of a Second Arctic Expedition to the Shores of the Polar Sea, in the years 1825, 1826, and 1827* (London, 1828).

Garry, Nicholas, "Diary of Nicholas Garry, Deputy-Governor of the Hudson's Bay Company from 1822-1835," John Bourinot, ed., *Proceedings and Transactions of the Royal Society of Canada*, Ottawa, 2nd ser., vol. VI (1900), sec. II, pp. 72-204.

Goldie, John, "In Memory of David Douglas," *British Columbia Historical Quarterly*, Victoria, B. C., II (1938), 89-94.

———, *Diary of a Journey in Upper Canada and some of the New England States, 1819* (Toronto, 1897).

Goodrich, Samuel G. (pseud.: Peter Parley), *Tales about Plants* (London, 1839).

Grant, W. C., "Remarks on Vancouver Island," *Journal of the Royal Geographic Society*, London, XXXI (1861), 208-213.

Guillet, Edwin C., *Early Life in Upper Canada* (Toronto: Ontario Publishing Co., Ltd., 1933).

Guthrie, John D., article in the *Oregonian,* Portland, Ore., July 14, 1932.

Hall, F. S., "Studies in the History of Ornithology in the State of Washington (1792-1932)," *The Murrelet,* Seattle, Wash., XV (1934), 3-16.

Hansen, Carl, "Pinetum Danicum," *Journal of the Royal Horticultural Society,* London, XIV (1892), 257-480.

Harvey, A. G., "David Douglas in British Columbia," *British Columbia Historical Quarterly,* Victoria, B. C., IV (1940), 221-243.

———, "Meredith Gairdner: Doctor of Medicine," *British Columbia Historical Quarterly,* Victoria, B. C., IX (1945).

———, "The Mystery of Mount Robson," *British Columbia Historical Quarterly,* Victoria, B. C., I (1937), 207-226.

Haskin, Leslie L., "David Douglas," *Oregon Sunday Journal,* Portland, Ore., April 5, 1925, sec. 4, p. 13.

Hedrick, Ulysses Prentiss, *A History of Agriculture in the State of New York* (Albany: New York State Agricultural Society, 1933).

———, *The Peaches of New York* (Albany, 1917).

———, *The Plums of New York* (Albany, 1911).

Hines, Gustavas, *Wild Life in Oregon* (New York, 1887).

Hitchcock, Charles H., *Hawaii and Its Volcanoes* (Honolulu, 1909).

Hooker, J. D., "A Sketch of the Life and Labours of Sir William Jackson Hooker," *Annals of Botany,* London, XVI (1902), ix-ccxxi.

Hooker, W. J., "A Brief Memoir of the Life of Mr. David Douglas, with Extracts from His Letters," originally published in the *Companion to the Botanical Magazine,* London, II (1836), 79-182, reprinted in the *Hawaiian Spectator,* Honolulu, II (1839), and in the *Oregon Historical Quarterly,* Portland, Ore., V and VI (1904 and 1905). Volume and page references in the footnotes refer to the *Oregon Historical Quarterly.*

———, *Flora Boreali-Americana; or, the Botany of the Northern Parts of British America,* 2 vol. (London, 1840), originally issued in twelve parts, the first in 1829.

———, *Icones Plantarum* (London, 1837).

———, "On the Botany of America," *Brewster's Edinburgh Journal of Science,* Edinburgh, III (1825), reprinted in the *American Journal of Science and Arts,* New Haven, IX (1825).

Hopkins, Flora Ball, *see* John Ball.

Howay, F. W., *see* Angus McDonald, E. O. S. Scholefield.

Howell, John Thomas, "A Collection of Douglas' Western American Plants," *Leaflets of Western Botany,* San Francisco, II (1937-1939), 59-62, 74-77, 94-96, 116-119, 139-144, 170-174, 189-192.

Howison, John, *Sketches of Upper Canada* (Edinburgh, 2nd ed., 1822).

Hunter, Thomas, *Woods, Forests, and Estates of Perthshire* (Perth, 1883).

Interprovincial Boundary Survey Commission, *Report of the Commission Appointed to Delimit the Boundary between the Provinces of Alberta and British Columbia,* part II (Ottawa, 1924).

Irving, Washington, *Astoria,* 2 vol. (Philadelphia, 1836).

———, *Adventures of Captain Bonneville* (1843).

Jackson, B. Daydon, *George Bentham,* English Men of Science Series, (London and New York, 1906).

James, George Wharton, *In and Out of the Old Missions of California* (Boston, 2nd ed., 1927).

Jarves, James J., *Scenes and Scenery in the Sandwich Islands* (Boston, 1843).

Jepson, Willis Linn, "David Douglas in California," *Madrono,* Berkeley, Cal., II (1933).

———, *A Flora of California* (San Francisco: Cunningham, Curtiss and Welch, 1936).

———, *The Silva of California,* Memoirs of the University of California, II (Berkeley, Cal., 1910).

Kane, Paul, *Wanderings of an Artist among the Indians of North America* (London, 1859).

Kip, Lawrence, "The Indian Council at Walla Walla, May and June, 1855," *Sources of the History of Oregon* (Eugene, Ore., 1897), vol. I, part 2.

Knight, Thomas Andrew, *A Selection from the Physiological and Horticultural Papers Published in the Transactions of the Royal and Horticultural Societies by the late Thomas Andrew Knight, Esq., . . . To which is Prefixed a Sketch of his Life,* Frances Acton, ed. (London, 1841).

Lambert, Aylmer Bourke, *A Description of the Genus Pinus* (London, 1803); 2nd ed., 2 vol. (London, 1828); 3rd ed., 2 vol. (London, 1832).

Lewis, Meriwether, and William Clark, *The Original Journals of the Lewis and Clark Expedition,* R. G. Thwaites, ed. (New York, 1904-1905).

Lewis, William S., *see* Angus McDonald, Ranald MacDonald.

Lindley, John, "Ribes Sanguineum," *Edwards's Botanical Register,* London, XVI (1830-31), plate 1349.

Lock, Alfred G., *Gold: Its Occurrence and Extraction* (London, 1882).

Lyman, Chester S., *Around the Horn to the Sandwich Islands and California, 1845-1850* (New Haven, 1924).

Lyman, George D., "The Scalpel under Three Flags in California," *California Historical Society Quarterly,* San Francisco, IV (1925).

Lyman, Henry M., *Hawaiian Yesterdays* (Chicago, 1906).

McDonald, Angus, "Angus McDonald: A Few Items of the West," F. W. Howay, Wm. S. Lewis, and Jacob A. Meyers, editors, *Washington Historical Quarterly,* Seattle, VIII (1917).

McDonald, Archibald, *Peace River. A Canoe Voyage from Hudson's Bay to the Pacific by the late Sir George Simpson (Governor, Hon. Hudson's Bay Company) in 1828. Journal of the late Chief Factor, Archibald McDonald,* Malcolm McLeod, ed. (Ottawa, 1872).

MacDonald, Ranald, *Ranald MacDonald, the Narrative of his Early Life on the Columbia under the Hudson's Bay Company's Regime,* Wm. S. Lewis and Naojiro Murakami, editors (Spokane, Wash., 1923).

MacKenzie, Cecil W., *Donald MacKenzie, "King of the Northwest"* (Los Angeles: I. Deach, Jr., 1937).

McLeod, Malcolm, *see* Archibald McDonald.

McLoughlin, John, *The Letters of John McLoughlin from Fort Vancouver to the Governor and Committee,* First Series, 1825-38, Edwin E. Rich, ed., Hudson's Bay Record Society Series (London: The Champlain Society, 1941).

Markham, Violet R., *Paxton and the Bachelor Duke* (London: Hodder and Stoughton, Ltd., 1935).

Mayne, Richard C., *Four Years in British Columbia and Vancouver Island* (London, 1862).

———, "Report on a Journey in British Columbia in the Districts bordering on the Thompson, Fraser, and Harrison Rivers," *Journal of the Royal Geographic Society,* London, XXXI (1861), 213-223.

Mazamas (a mountaineering club), *Mount Hood* (Portland, 1920).

Meany, Edmond S., "Origin of Washington Geographical Names," *Washington Historical Quarterly,* Seattle, XIV (1923).

———, *see* Charles Wilkes.

Meisel, Max, *Bibliography of American Natural History,* 2 vol. (Brooklyn, N. Y., 1926).

Menzies, Archibald, *Hawaii Nei 128 Years Ago,* W. F. Wilson, ed. (Honolulu, 1920).

———, *Menzies' Journal of Vancouver's Voyage, April to October, 1792,* C. F. Newcombe, ed. (Victoria, B. C., 1923).

Merk, Frederick, *Fur Trade and Empire; George Simpson's Journal: . . . 1824-1825* (Cambridge, Mass.: Harvard University Press, 1931).

Meyers, J. A., "Finan McDonald—Explorer, Fur Trader and Legislator," *Washington Historical Quarterly,* Seattle, XIII (1922), 196-208.

Michaux, André, *Flora Boreali-Americana,* 2 vol. (Paris, 1803).

Morice, A. G., *History of the Northern Interior of British Columbia* (Toronto, 3rd edition, 1905).

Murakami, Naojiro, *see* Ranald MacDonald.

Murray, J. M., "David Douglas," *Forestry,* London, V (1931), 154-158.

Newcombe, C. F., *see* Archibald Menzies.

New York Horticultural Society, Constitution and By-laws (New York City, 1824).

Norris, Leonard, "Some Place Names," *Sixth Report of the Okanagan Historical Society, 1935* (Vancouver, B. C., 1936), pp. 133-158.

Nuttall, Thomas, *A Journal of Travels into the Arkansas Territory during the Year 1819* (Philadelphia, 1821).

———, "A Catalogue of a Collection of Plants made chiefly in the Valleys of the Rocky Mountains or Northern Andes, towards the sources of the Columbia River, by Mr. Nathaniel B. Wyeth, and described by T. Nuttall," *Journal of the Academy of Natural Science of Philadelphia,* VII (1834), 5-60.

———, *The Genera of North American Plants,* 2 vol. (Philadelphia, 1817).

Ogden, P. S., "Peter Skene Ogden's Notes on Western Caledonia," W. N. Sage, ed., *British Columbia Historical Quarterly,* Victoria, B. C., I (1937).

Oliver, E. H., ed., *The Canadian Northwest, Its Early Development and Legislative Records,* 2 vol. (Ottawa, Ont., 1914-15).

Oliver, F. W., ed., *Makers of British Botany* (Cambridge, England, 1913).

Palliser, John, *Journals, Detailed Reports, and Observations* (London, 1863).

Parker, Samuel, *Journal of an Exploring Tour beyond the Rocky Mountains . . .* (Ithica, N. Y., 4th edition, 1844).

Parley, Peter (pseud.), *see* S. G. Goodrich.

Powers, Kate N. B., *see* John Ball.

Praeger, R. Lloyd, "William Henry Harvey, 1811-1866," in *Makers of British Botany,* F. W. Oliver, ed. (Cambridge, England, 1913).

Pursh, Frederick, *Flora Americae Septentrionalis,* 2 vol. (London, 1813).

Ravenscroft, Edward, ed., *The Pinetum Brittannicum,* 3 vol. (London and Edinburgh, 1884).

Reed, Henry E., "William Johnson," *Oregon Historical Quarterly,* Portland, Ore., XXXIV (1933), 314-323.

Rich, Edwin E., *see* John McLoughlin.

Richardson, John, *Arctic Searching Expedition,* 2 vol. (London, 1851).

———, *Fauna Boreali-Americana, or the Zoology of the Northern Parts of British America,* 2 vol. (London, 1829, 1831).

———, "Report on North American Zoology," *Report of the Sixth Meeting of the British Association for Advancement of Science, 1836,* V (London, 1837), pp. 121-224.

Rickard, T. A., "Indian Participation in the Gold Discoveries," *British Columbia Historical Quarterly,* Victoria, B. C., II (1938), 3-18.

Ross, Alexander, *The Fur Hunters of the Far West,* 2 vol. (London, 1855).

Russell, Israel C., *Geology and Water Resources of Nez Perce County, Idaho,* Water Supply and Irrigation Papers, United States Geological Survey, no. 53 (Washington, D. C., 1901).

Sabine, Edward, "Report on the Variations of Magnetic Intensity observed at different Points of the Earth's Surface," *Report of the Seventh*

Meeting of the British Association for the Advancement of Science (Report for 1837), VI (London, 1838), 1-85, 500.
Sage, W. N., *see* Peter S. Ogden.
Saunders, Charles F., *Western Wild Flowers and Their Stories* (Garden City, N. Y. : Doubleday, Doran & Co., Inc., 1933).
Scholefield, E. O. S., and F. W. Howay, *British Columbia,* 4 vol. (Vancouver and Chicago, 1914).
Scouler, John, "Dr. John Scouler's Journal of a Voyage to N. W. America," *Oregon Historical Quarterly,* Portland, Ore., VI (1905), 54-75, 159-205, 276-287.
Simpson, George, *see* Frederick Merk.
Smet, Pierre Jean de, *Oregon Missions and Travels over the Rocky Mountains in 1845-46* (New York, 1847).
Somerville, Thomas, "An Early Hero of the Pacific; Douglas, the Botanist," *Overland Monthly,* San Francisco, VII (August, 1871).
Stenhouse, John H., "Birds of Historic Interest in the Royal Scottish Museum. Birds of Franklin's Overland Expeditions," *Scottish Naturalist,* Edinburgh, May-June 1930, pp. 75-84.
Stutfield, Hugh E. M., and J. Norman Collie, *Climbs and Explorations in the Canadian Rockies* (London, New York, and Bombay, 1903).
Sudworth, George B., *Forest Trees of the Pacific Slope* (Washington, D. C.: United States Depertment of Agriculture, Forest Service, 1908).
Sullivan, Maurice S., *The Travels of Jedediah Smith* (Santa Ana, Calif.: The Fine Arts Press, 1934).
Sweetser, Albert R., articles in *Sunday Oregonian,* Portland, Ore., sec. 5, p. 11, April 12, 1925, and sec. 5, p. 5, November 7, 1926.
Symons, T. W., *Report on an Examination of the Upper Columbia River* (Washington, D. C., 1882).
Thompson, David, *David Thompson's Narrative of his Explorations in Western America, 1784-1812,* J. B. Tyrrell, ed. (Toronto, 1916).
Thorington, J. M., "The Centenary of David Douglas' Ascent of Mount Brown," *Canadian Alpine Journal,* Banff, Alberta, XVI (1926-27), 185-197.
———, *The Glittering Mountains of Canada* (Philadelphia, 1925).
Thwaites, R. G., *see* Meriwether Lewis.
Tolmie, S. F., "My Father: William Fraser Tolmie," *British Columbia Historical Quarterly,* Victoria, B. C., I (1937), 227-240.
Townsend, John K., *Narrative of a Journey across the Rocky Mountains to the Columbia River and a Visit to the Sandwich Islands, Chili, etc.* (Philadelphia, 1839).
Tyrrell, J. B., *see* David Thompson.
Urquhart, J. D., *Historic Sketches of Scone* (Perth, 1883).
Wheeler, Arthur O., "Mounts Brown and Hooker," *Canadian Alpine Journal,* Banff, Alberta, XVII (1928), 66-68.

White, James, *see* Edward Ermatinger.
Wilkes, Charles, "Diary of Wilkes in the Northwest," Edmond S. Meany, ed., *Washington Historical Quarterly,* Seattle, XVI (1925).
———, *Narrative of the United States Exploring Expedition. During the years 1838, 1839, 1840, 1841, 1842,* 5 vol. (Philadelphia, 1845).
Wilson, James (1795-1856), *Illustrations of Zoology* (Edinburgh and London, 1831).
Wilson, William F., *David Douglas, Botanist, at Hawaii* (Honolulu, 1919).
———, *see* Archibald Menzies.
Work, John, "The Journal of John Work," T. C. Elliott, ed., *Washington Historical Quarterly,* Seattle, installments as follows: journal for November and December, 1824, in III (1912); for December 15, 1825, to June 12, 1826, in V (1914); for July 5 to September 15, 1826, in VI (1915).
Wyeth, Nathaniel J., "The Correspondence and Journals of Captain Nathaniel J. Wyeth, 1831-6," F. G. Young, ed., *Sources of the History of Oregon,* I (Eugene, Ore., 1899), parts III and VI.
Young, F. G., *see* Nathaniel J. Wyeth.

PERIODICALS AND PROCEEDINGS OF LEARNED SOCIETIES

Academy of Natural Science, *Journal,* Philadelphia.
Academy of Sciences of St. Louis, *Transactions.*
American Journal of Science and Arts, New Haven, Conn.
Annals of Botany, London.
Botanical Gazette, Indianapolis, Ind.
Botanical Magazine, London.
Brewster's Edinburgh Journal of Science, Edinburgh.
British Association for the Advancement of Science, *Report,* London.
British Columbia Historical Quarterly, Victoria, B. C.
California Historical Society Quarterly, San Francisco.
Canadian Alpine Journal, Banff, Alberta.
Canadian Naturalist and Geologist, Montreal.
Century Magazine, New York.
Companion to the Botanical Magazine, London.
Cottage Gardener, London.
Curtis's Botanical Magazine, London.
Edinburgh Journal of Science, Edinburgh.
Edinburgh Philosophical Journal, Edinburgh.
Edwards's Botanical Register, London.
Forestry, London.
Gardener's Chronicle, London.
Gardener's Magazine (Loudon's), London.
Gentleman's Magazine, London.
Geological Society, *Proceedings,* London.

Geological Society, *Transactions,* London.
Hawaiian Spectator, Honolulu.
Hilo Tribune, Hilo, Hawaii.
Hooker's Botanical Miscellany, London.
Honolulu Star-Bulletin.
Horticultural Society of London, *Transactions.*
Horticultural Society, Royal, *Journal,* London.
Hutchings' Illustrated California Magazine, San Francisco.
Ke Kumu Hawaii (newspaper), Honolulu.
Leaflets of Western Botany, San Francisco.
Linnean Society, *Transactions,* London.
Madrono (Journal of California Botanical Society), Berkeley, Cal.
Magazine of Natural History (Loudon's), London.
Murrelet, Seattle, Wash.
Okanagan Historical Society, *Report,* Vancouver.
Oregon Historical Quarterly, Portland, Ore.
Oregon Pioneer Association, *Transactions,* Salem, Ore.
Oregon Sunday Journal, Portland.
Oregonian, Portland, Ore.
Overland Monthly, San Francisco.
Pacific Historical Review, Los Angeles.
Paradise of the Pacific, Honolulu.
Paxton's Magazine of Botany, London.
Polynesian, Honolulu.
Quarterly Journal of Science, London.
Quarterly Review, London.
Royal Geographic Society, *Journal,* London.
Royal Geographic Society, *Proceedings,* London.
Royal Horticultural Society, *see* Horticultural Society.
Royal Society, *Transactions,* London.
Royal Society of Canada, *Proceedings and Transactions,* Ottawa, Ont.
Sandwich Island Gazette, Honolulu.
Washington Historical Quarterly (now *Pacific Northwest Quarterly*), Seattle, Wash.
Zoological Journal, London.
Zoological Society, *Proceedings,* London.

MANUSCRIPTS

L–letter(s); J–journal(s). Figures refer to the following depositories:

1. Bancroft Library, University of California, Berkeley, California.
2. British Museum, London.
3. Columbia University Library, New York City.
4. Dominion Archives, Ottawa, Ontario.
5. Hawaiian Mission Children's Society Library, Honolulu.

6. Hudson's Bay Company Archives, London.
7. Royal Botanic Gardens, Kew, Surrey, England ("Kew Gardens").
8. Linnean Society, London.
9. Peabody Museum, Salem, Massachusetts.
10. Provincial Archives, Victoria, British Columbia.
11. Public Record Office, London.
12. Royal Horticultural Society, London.
13. Royal Society, London.

Anderson, Alexander Caulfield, "History of the Northwest Coast" .. 1
Bentham, George, Autobiography; Diary 7
Booth, William Beattie, Memorandum on David Douglas 7
California Departmental Records 1
Chamberlain, Levi, J 5
Clinton, DeWitt, L to Joseph Sabine (drafts) 3
——— Memorandum, "Horticultural Society of London" 3
Colonial Office Papers (British Government) 11
Coulter, Thomas, L to W. J. Hooker 7
Douglas, David, Book of Field Sketches, Fort Okanagan to Quesnel River, 1833 10
——— L to DeWitt Clinton 3
——— L to Gov. Figueroa; L to W. E. P. Hartnell 1
——— L to R. W. Hay (Colonial Office) 11
——— J; and L to Joseph Sabine; L to N. Vigors 12
——— L to Joseph Sabine 13
——— L to William Fraser Tolmie 10
——— L to W. J. Hooker 7
Douglas, John, "Biographical Notes of David Douglas" 7
Gairdner, Meredith, L to W. J. Hooker 7
Garry, Nicholas, L to R. W. Hay (Colonial Office) 11
Hay, R. W., L to Barrow (Admiralty) 11
Hooker, William Jackson, L to Dr. John Richardson 7
Horticultural Society of London, Minutes of Council 12
Hudson's Bay Company, Account Books; L to Horticultural Society of London 6
Linnean Society, Minutes of Council 8
Lyman, Mrs. D. B., J 5
McDonald, Archibald, L to Edward Ermatinger 4
——— L to W. J. Hooker 7
McLeod, John, J 4
Monterey Company of Foreigners, Orderly Book 1
Reynolds, Stephen, J 9
Richardson, Dr. John, L to W. J. Hooker 7
Roberts, George B., "Recollections of the Hudson's Bay Company" .. 1
Rooke, T. C. R., L to George Bentham (copy) 7

Sabine, Edward, L to W. J. Hooker 7
——— "Observations taken on the Western Coast of North America by the late Mr. David Douglas, with a Report on his Papers," 1837 13
Sabine, Joseph, L to Robert Brown 2
——— L to DeWitt Clinton 3
——— L to David Douglas (copies) 12
Simpson, Aemilius, 1826 Journal 6
Tod, John, "New Caledonia" 1
Tolmie, William Fraser, L to W. J. Hooker 7
Young, T. C., "Extracts from George McDougall's Journal" 10

INDEX

(HBC *means* Hudson's Bay Company. *The contents of* APPENDIX B, pp. 254–260, *are not indexed.*)

Aberdeen, Wash., 64
Abies amabilis, 58–60
bracteata, see *venusta*
grandis, 171; size, 196
nobilis, 58–60; seeds, plants, 171–172
venusta, 183–184
Acacia koa, use, 215
Acid, fluoric, discovery denied, 224n–225n
Admiralty, British, map of Douglas's travels, 133, 152
Albany, N.Y., 27, 31–32
Albatross, 45
Amherstburg, Upper Canada, 28–31
Anderson, William, botanist with Captain Cook, 40
Andersonian Institution, Glasgow, receives birds from Douglas, 246
Animals, *see* names
Ann Maria, voyage to U.S.A., 22
Annance, François Noel, HBC, 107
Apple trees, 105–106
Arbutus, found by Menzies, 40; introduced by Douglas, 55, 154
Arguella, Dona Concepción, 189
Argyroxiphium sandwicense, seeds, 216
Arrow Lakes, Columbia River, 109
Arrowhead, roots as food, 63
Arrowsmith's maps, altitudes of mounts Brown and Hooker, 140–141
Artemisia, 71, 86
Assiniboine, Fort, 119–120
Astoria, *see* Fort George
Astoria, book, 134–136, 140
Athabaska Pass, 112–117; height, 130–143
Athabaska River, 116, 118–120
Atkinson, ——, Douglas's sister, 8
Atkinson, William, English architect, 8, 155
Audibertia incana, 185

Baby-blue-eyes, 176
Back, George, Arctic explorer, 128–129
Baker's Bay, 47
Ball, John, Fort Vancouver school teacher, 195, 212
Bandinel, James, Foreign Office clerk, 237
Banks, Peter, English gardener, 110n, 128
Barnston, George, HBC, 105, 139; quoted, 106, 168, 171n; tribute, 239
Barton, William P. C., founder of University of Pennsylvania Garden, 25
Bartram, John, founder of America's oldest botanic garden, 33–34
Batavia, N.Y., 28
Bear grass, 58
Beattie, William, Scone Palace gardener, 11–14
Beaver, first steamship in Pacific Northwest, 193n
Beechey, Captain F. W., collects plants for Hooker, 152; criticism of California missionaries answered by Douglas, 188; Douglas's plants described in *Botany of Captain Beechey's Voyage,* 186
Belcher, Captain Sir Edward, 167; quoted, 174

Bentham, George, quoted, 146; named secretary of Horticultural Society, 191–192; herbarium, 240n
Berberis aquifolium, 36; seeds, 90
Big Bend, Columbia River, 111
Big Eddy, Columbia River, 109
Big Hill, 113, 115–116
Billy, *see* Dog
Birds, *see* names
Bitter root, 74
Black, Samuel, HBC, at Fort Walla Walla, 71, 79; at Fort Kamloops, quarrel with Douglas, 202–203, 205
Blazing-star, 55, 184–185
Blue Mountains of Oregon, journeys into (1826), 79–82, 84–85, 112, the first ascent, 246; (1830), 168; (1833), 205
Bluebell Mine, Kootenay Lake, 210–211
Boat Encampment, Columbia River, 111–113
Bois Blanc Island, Detroit River, 29
Boki, Madame, Honolulu, 166
Bolabola, 252–253
Bonaparte, Joseph, Bordentown, New Jersey, 26
Booth, William Beattie, gardener, Horticultural Society of London, 148, 150
Bordentown, N.J., 26
Botanical Register, see *Edwards's Botanical Register*
Botany of Captain Beechey's Voyage, 186
Brackenridge, J. D., investigates Douglas's death, 252
Brenchley, Rev. Julius L., provides tombstone, 241
Brewster, Wash., 72
Briscoe, Henry, 28–30
British Columbia, northern interior, *see* New Caledonia
British Museum, London, Douglas's rocks, 211; herbaria, 227n, 240n
Brodiaea grandiflora, 55, 184–185
Brookes, Joshua, English anatomist, 38
Brown, J., of Perth Nursery, 13, 16
Brown, R., of Perth Nursery, 13, 16
Brown, Robert, British botanist, mountain named after, 132; *Paeonia Brownii*, 143
Bruce, Billy, HBC gardener, 167
Buccleuch, Duke of, 8
Buchier, ——, missionary at Red River, 126
Buel, Jesse, Albany, N.Y., 31
Buffalo, N.Y., 28, 31
Buffalo hunt, North Saskatchewan River, 122–123
Bullock, kills Douglas, 233–238, 251–253
Burlington, N.J., 34

Cabot, Rev. Pedro, California missionary, 183, 188
California, Douglas travels in, 172–189, 212–213; passport, 174, 175; itinerary (1831), 176; climate, 179–180; gold found, 209–210; missionaries defended, 187–188; women praised, 188–189
Calipio, Hawaiian guide, 223–226
Calochortus, 54, 184
Camassia esculenta, roots as food, 74, 103; flatulence from, 56
Cambridge University herbaria, 227n, 240n
Camino Real, El, old California road, 176, 180
Camphor weed, 94
Canada, Upper, Douglas's travels into, 28–31
Canadian Pacific Railway, 136
Candolle, Augustin Pyramus de, Swiss botanist, 180; his *Prodromus* describes Douglas's plants, 186
Cape Disappointment, 63
Cape Horn, 45
Cape Shoalwater, 63
Captain Bonneville, book, 136
Cardinal, Jacques, packer, Jasper region, 118
Cartier, Jacques, French explorer, 20
Cascade Mountains, 58, 60, 246
Cascades, Columbia River, 57, 89, 169
Castanopsis chrysophylla, 95–96
Castle, ——, sawmill owner, Hawaii, 216
Celilo, *see* Great Falls

Centrenose, Chief, Umpqua River, 98
Cervus leucurus, 148
Chancellor Livingstone, Hudson River steamboat, 32
Charlton, Richard, British Consul, Honolulu, 193, 214; investigates Douglas's death, 234–237, 251; wife receives letter from Douglas, 228–231
Chehalis River, 64–65, 111
Chestnut, golden-leafed, 95–96
Chironia, 24
Chumtalia, Chief, Columbia River, 58, 89, 169; death, 170
Clark, William, hunter, Juan Fernandez Island, 45–46
Clarkia pulchella, 54
Clethra, 96
Clinton, DeWitt, ex-governor of New York, assists Douglas, 27–28, 31, 33, 36; praises Douglas 38
Clover, 81, 86
Coan, Rev. Titus, Hilo, Hawaii, 252, 253n
Cochran, Rev. William, Red River Colony, 126
Cock de Lard, guide, Blue Mountains, 84–85
Cockqua, Chief, Douglas entertained by, 52–53; pot luck with, 105
Cockroaches, 192
Coleman, Arthur Philemon, makes ascent of Mount Brown in 1893, 136–137
Collie, Alexander, 175
Collinsia grandiflora, 55
Colocasia antiquorum, 218–219
Colnett, Captain James, on Northwest Coast, 40
Colonial Office, London, assistance from, 161, 163; receives news of Douglas's death, 238
Colorado River, visit abandoned, 190; visited by Coulter, 180, 181
Columbia River, bar at mouth, 46-47; travels on, *see* Douglas, David, Journeys
Committee's Punch Bowl, Athabaska Pass, 116–117
Concomly, Chief, host, 62–63; father-in-law of Archibald McDonald, HBC, and grandfather of Ranald, 195
Condor, California, specimens, 96, 106; given to Zoological Society, 147–148; search for nest, 181
Connolly, William, HBC New Caledonia brigade, 76–77, 83
Connor, Ralph, author of *The Man from Glengarry,* 123
Cook, Captain James, on Northwest Coast, 40; on height of Mauna Loa, 224
Cooper, Juan Bautista Roger, California pioneer, 183
Coquille River, 102–103
Corney, Captain Peter, *Eagle,* 237
Coulter, Thomas, British botanist, 180–182, 184
Cowlitz River, 65, 111
Cox, Ross, altitude of Rocky Mountains, 140
Coxe, William, Burlington, N.J., 34
Creamcups, 184–185
Currant, red-flowered, seeds, 90; great novelty, 146

Dalles des Morts, Columbia River, 110
David Douglas, Mount, Cascade Range, Ore., 143n
Davidson, Captain, *Dryad,* 90
Davis, ——, Nedo Ranch, Hawaii, 233, 236
Davis, William Heath, attended by "Doctor" Douglas, 189
Dawson, Dr. George M., names Mount Douglas, 143
Dease, John Warren, HBC, accompanied to Kettle Falls, 72–73; makes return arrangements, 86–87
Dease, Peter Warren, HBC chief factor, Fort St. James, 203
Deer, new species, 108, 148
Dendromecon rigidum, 184–185
Deppe, Ferdinand, German naturalist, 181
Description of the Genus Pinus, Douglas fir, 47, 154; redwood tree, 177
De Smet, *see* Smet

Detroit, not visited but river crossed to botanize, 30
Dick, William, Pennsylvania University Garden, 25, 33
Dickson, John, Rio de Janeiro, 43
Diell, Rev. John, Honolulu, plans Kilauea visit with Douglas, 232; Douglas's death, 234–236
Doak, Thomas, California pioneer, 183
Dog "Billy," described, 172, 201; on Douglas's death, 233–234, 253; sent to England, 237
Douglas, David, birth, 7; youth, 7–16; parents, 7; lack of self control, 8–10, 151, 157–161; education, 8–16, 34, 161–163; journeys: U.S. and Upper Canada, 19–39, Northwest America (1824–1827), 3, 40–147 (Gray's Harbor, 62–64, 111, Hudson Bay, 107–129, Umpqua, *see* Pine, sugar), list and distances, 111–112, 128, cost, 146, Hooker's map, 132–134, 152; (1829–1830), 160–173; (1832–1833), 193–208, 210–211 (New Caledonia, 197–206, 210); California, 172–189, 209–210, 212–213, Sandwich Islands (1830), 166, (1832) 190–193, (1833–1834), 212–237, 240–241, 251–253; *see also* Blue Mountains, Fort Colville *and other place names;* illness in U.S.A., 32, Fort Vancouver (1825), 62, (1830), 171, Fort Colville, 75, Hudson Bay, 129, London, 146–147, 157, 164; modesty, 39, 124, 146; character, 42–43; suffers from shortage of food, 52, 53, 63–64, 74, 79–80, 83, 87–89, 93–94, 97, 103, 105, 119–121, 147; known as "Grass Man" by Indians, 55, 63, 70, 83, 88; impaired eyesight, 69, 77, 82, 164, 180, 200, 253; mountains named after Douglas, 143; fame and honors, 144–160; salary, 161; observations to determine latitude and longitude, instruction in, 161–162, made on ship, 165, at Oahu, 166, at Fort Vancouver, 167, 195–196, in California, 178, in New Caledonia, 201, 204, 206, no remuneration for, 240n, success at, 246; termed "Doctor" Douglas, 189, *see also* 122–123; resignation from Horticultural Society, 191; field sketches, 202–203; quarrel with Samuel Black, HBC, 202–203; wreck on Fraser River, 204–206; death, 233–238, accident or murder, 251–253; tributes to, 235–236, 238–247; memorials, 239–240, 241, plants, 241–253, Douglas fir, 241–243; successors, 245
Douglas, George, David's brother, 8
Douglas, James, HBC, 70, 76; on height of Rocky mountains, 135
Douglas, John, David's father, 7–9; death, 124
Douglas, John, David's brother, 8; on David's inquisitiveness, 10–11; receives news of David's death, 238
Douglas fir, *see* Fir
Douglasia nivalis, 117
Drummond, Jean, Douglas's mother, 7
Drummond, Thomas, plant collector with Franklin's Second Arctic Expedition, 120–121, 123; escape in Hudson Bay, 128–129; on height of Rocky Mountains, 140; Hooker suggests California for him and Douglas, 160
Dryad, takes plants to England, 90; voyage to California, 172; voyage to Honolulu, 212–214
Ducks, seasick, 36
Duel, Douglas challenged by Samuel Black, HBC, 202–203
Duran, Rev. Narciso, California missionary, 188

Eagle, bald, 127–128; golden, 121–122, 127–128
Eagle, vessel, 161, 237
Edinburgh University Museum, birds, 246
Edwards's Botanical Register, features Douglas plants, 149

Elerslie, N.Y., 27
Elgin Botanical Garden, New York City, 24
Elk Creek, sugar pine search, 95–97
Elkton, Ore., 96–97
Engelmann, Dr. George, American botanist, 59–60
Epigea repens, 37–38
Erie Canal, 27–28, 31
Eriogonum, 86
Eriophyllum lanatum, 244–245
Ermatinger, Edward, commands HBC express, 107–124
Ermatinger, Francis, HBC, 76
Erythronium grandiflorum, 73; roots, 108

Fairy lantern, 184
Fannaux, ——, HBC, 103
Farewell-to-spring, 244
Fauna Boreali-Americana, specimens for, 153, 172
Fern, giant pala, 227
Ferns, collected in Hawaii, 217–218, 227
Fever epidemic in Pacific Northwest, 1830–1833, 170–172, 200, 206–207
Figueroa, Jose, governor of California, 212–213
Finlay, Jacques Raphael, trapper, Spokane House, meagre fare, 73–74; better fare, 86
Finlayson, Duncan, HBC, 212
Fir, nomenclature, 58–59
  bristlecone or charming, 183–184
  Douglas, 57–59; espied, 3, 47; seeds, 90, 145; name, 57–59, 154, 242–243; qualities, 57–58, 242; plants sold, 191
  lovely, 58–60
  noble, 58–60; seeds, plants, 171–172
  Santa Lucia, 183–184
  white, 171; size, 196
Fires, forest, 93
Fischer, Dr. Gotthelf, St. Petersburg, 198
Fish, ——, a friend, Douglas's character, 9–10
Flora Boreali-Americana (Michaux), 19–20; (Hooker), Mount Brown and Mount Hooker, 132–134; Douglas's plants, 152, 156, 184
Floy, Michael, N.Y., 32, 37
Fort Alexander, Winnipeg River, 124
Fort Alexandria, Fraser River, 203, 205
Fort Carlton, North Saskatchewan River, 119, 123
Fort Colville, Douglas's visits: (1826), 68–75, 86–87, 112, 211; (1827), 108
Fort Edmonton, 119–122
Fort Fraser, 203
Fort Garry, Red River Colony, 125–126
Fort George (Astoria), 48–49, 51, 171, 172
Fort George Canyon, Fraser River, 204–205
Fort Hope, Fraser River, 59–60
Fort Kamloops, 202–203, 205
Fort Langley, 105
Fort Nisqually, 196
Fort Okanagan, visits: (1826), 71–72, 76, 87–88, 112; (1833), 200–201, 205
Fort Ross, Calif., 178
Fort St. James, 119, 199, 201, 203–204
Fort Umpqua, 96
Fort Vancouver, 49, 51, 60–62, 65–66, 89; distances covered, 111–112; growth (1830), 166–167, (1832), 193–195; longitude, 167, 195, 196
Fort Walla Walla, Douglas's visits: (1826), 71, 76–79, 83, 88; (1830), 167–169; (1833), 205
Franklin, Benjamin, 25
Franklin, Captain John, on Arctic expedition, 42, 120, 123, 128; accompanies Douglas up Lake Winnipeg, 124; has grouse named after him, 116, 153–154
Fraser River, journey, 201, 203–205; wrecked, 204–205
Fritillaria pudica, 71; bulbs, 108

Gabilan Range, Calif., 176
Gairdner, Dr. Meredith, HBC, 207–208; tribute by, 239, 243; on

mystery of Douglas's death, 251–252
Galapagos Islands, *see* James Island
Garry, Nicholas, HBC deputy governor, London, orders animal skins, 108; escapes from drowning, 127; discusses Oregon boundary, 156
Garry, Spokane, Indian boy, 125–126
Garrya elliptica, 185
Gaultheria shallon, *see* Salal
Geese, Sandwich Island, 192
Genus Pinus, A Description of the, Douglas fir, 47, 154; redwood tree, 177
Geographic Board of Canada, conclusions on Mount Brown and Mount Hooker, 137–138
Geological Society of London, specimens for, 148, 211
Geology, *see* Minerals
George IV, Horticultural Society patron, 18; delinquent, 191
Gesneria Douglasii, 44–45
Gila River region, visited by Coulter, 181
Gilia, 184
Glasgow Botanic Garden, Douglas employed at, 14; growth, 156
Godetia amoena, 244
Gold, discovery of, 209–210
Golden club, 24
Goldie, John, 20, 198
Goodrich, Rev. Joseph, host to Douglas, 214, 215, 217, 227; at time of Douglas's death, 234–236
Gooseberry, wild, Snake River, 85; California, 175
Grand Coulee, 72
Grande Ronde, 80, 81
Grant, Cuthbert, Warden of the Plains, Red River Colony, 126
Grape, Oregon, *see* Oregon grape
Grave, Captain, *Eagle*, 161
Gray's Harbor, 62–64, 111
Great Dalles, Columbia River, 54, 83, 88, 111
Great Falls, Columbia River, 51, 54, 70, 88, 107, 111
Greenwich Observatory, study at, 162–163
Groundhogs, sent to England, 82
Grouse, 76, 108; Franklin's, 116, 153–154
Gurney, Edward, hunter, Hawaii, suspected of murdering Douglas, 233–236, 251–252

Hall, ——, Hawaii, opinion on Douglas's death, 235, 236, 251
Hamlyn, Dr. Richard J., Douglas's traveling companion, Red River to Hudson Bay, 126–127
Hansen, Carl, reports re-discovery of *Abies amabilis*, 59–60
Hanwell, Captain Henry, *William and Ann*, 41, 62; gets Horticultural Society's medal, 48, 61
Harper, Rev. Theophilus, Red River Colony, 126
Harrison, Henry, Rio de Janeiro, 44
Harrison, William, Rio de Janeiro, 44
Hartnell, William Edward Petty, Monterey, Calif., Douglas resides with, 175, 182, 189, 213
Hartweg, Karl Theodor, 245
Hatshepsut, Egyptian queen, 17
Hawaiian Islands, *see* Douglas, journeys of
Hay, R. W., Colonial Office, grant for instruments, 163
Hayes River, boat smashed, 127
Hector, Dr. James, 136
Heliotrope, wild, 184–185
Highlands, Scottish, botanizing in, 13, 16
Hilo, base for journeys in Hawaii, 219, 226, 228, 233–237
Hilo Tribune, reports Douglas murdered, 252–253
Hoffman, Martin, New York merchant, 26–27
Hogg, Thomas, New York nurseryman, 23–26, 35
Honey, wild, 93
Honeysuckle, 29, 55–56
Honolii, John, Hawaiian guide, 215–226
Honolulu, Douglas visits: (1830), 166; (1832), 190–193; (1833), 214; (1834), 228–232, 235–237

Hooker, Dr. William Jackson, professor of botany, Glasgow, tells of Douglas's pet birds, 10; teaches Douglas botany, 14–16; recommends him to Horticultural Society, 18, 41; recommends Scouler, 42; writes *Flora Boreali-Americana*, 132–134, 152, 156, 184; develops herbarium, 152–153, 156; collects California plants, 181, 185–186, 190; receives Sandwich Island plants, 192, 226; receives plants from Tolmie, 208; is annoyed with Douglas, 160; sorrows at his death, 238–239; respected by Douglas, 153, 156, 243; interested in Douglas's journal, 149, 152
Hooker and Arnott, *Botany of Captain Beechey's Voyage*, 186
Horticultural Society of London, employs Douglas, 18–20, 40–41, 243; fate of recent collectors, 42; garden, 19, 37–38, 151; *Transactions* features newly discovered plants, 18, 37, 148, 186; satisfaction with Douglas, 37–38, 124, 144–146, 149; future plans, 151, 157, 159–161, 164; California plants, 185–186, 190; reorganization, 190–192; resignation of Douglas from, 190–192; hears of Douglas's death, 238; buys Douglas's Sandwich Island herbarium, 240n
Hosack, Dr. David, New York, 23–24, 26, 27, 33, 36, 39
Hound's tongue, plant, 74
Howell, John Thomas, quoted, 187
Howison, John, quoted, 25
Hudson Bay, journey to, 107–128; escape in the bay, 128–129
Hudson's Bay Company, assistance from, 41, 61, 146, 161; relations with Indians, 54; New Caledonia brigade, 76, 83, 201; Oregon boundary suggestions, 155–156; fever deaths, 171; trade goods, 172–173
Hudson's Bay Company express, Douglas's journeys with: (1825), 65–66; (1826), 69–72; (1827), 107–129
Hyacinth, wild, 55, 184–185

Illustrations of Zoology, 153, 246
Incense tree, 183–184
Indians, clothing, 48; war dance, 48; shooting, 53; food, 51, 52, 56, 58, 63, 74, 88, 89, 98–99, 105; filth, 64; barter, 84 (*see also* Tobacco); burials, 74; hostility, 54, 70, 81, 101, 102, 107, 207; quarrels, 52, 81, 85–87; polygamy, 99; potlatch, 169; names for Douglas, 54-55; gifts from Douglas, 170, 201; treatment by Douglas, 83, 88, 244
Interprovincial (Alberta-B.C.) Boundary Commission, Mount Brown and Mount Hooker, 137–138, 142

Jaggar, Dr. T. A., quoted, 224n–225n
James Island, Galapagos Group, plants and birds, 46
Jasper, Alta., 118
Jasper House, a stag dance, 118
Jasper Lake, altitude, 139
Jefferson, Thomas, president, 21, 26
Jefferson, Mount, *see* Mount Jefferson
Jefferson, Ore., 94
Jeffrey, John, plant collector, 59, 245
Jepson, Willis Linn, discusses sugar pine, 92, 104; discusses Douglas's California plants, 187
John, Diell's servant, 232, 235, 251
Johnson, William, HBC, Douglas's servant in the Pacific Northwest, 167–168, 196, 201–205
Jones, Rev. David T., Red River Colony, 126
Journal (Douglas's), publication, 149–150, 152, 261–262
Juan Fernandez Island, 45–46
Judd, Dr. Gerritt P., Honolulu, makes postmortem, 237

Kaluakauka, place of Douglas's death, 241
Kapapala, Hawaii, 222, 226
Kendall, Lt. E. N., escape in Hudson Bay, 128–129

Kennedy, John, HBC, Douglas's companion on sugar pine journey, 94, 103–104
Kettle Falls, *see* Fort Colville
Kettle River, first ascent, 75
Kew Gardens, Douglas's herbaria, 227n, 240n
Kilauea volcano, 219–222, 226–231
Kinlock, George, California pioneer, 183
Kinnikinnick berry, 63
Kinnoul, Perthshire, school, 9–11
Kipling, Captain Charles, *Dryad*, 212
Kittson, William, HBC, inspects Kettle River, 75
Klebnikoff, Cyril, 198
Knight, Thomas A., president of Horticultural Society of London, 18; on Douglas's shyness, 39; discusses Douglas's hardships, 144
Kohala Point, Hawaii, 232
Kootenay Lake, mineral discovery credited to Douglas, 210–211

Laframboise, Michel, HBC, Douglas's companion on sugar pine journey, 102
Lake Erie, 28, 31
Lake St. Clair, 29–30
Lake Waha, 84–85
Lake Winnipeg, 123–126
Lama, vessel, Honolulu to Fort George, 193
Lambert, Aylmer Bourke, 92. *See also Genus Pinus*
Landreth, David, Philadelphia nurseryman, 34
Latitude and longitude, *see* Douglas, David, Observations
Laurel, California, seeds, 95–96
Lesser Slave River, 119
L'Etang, Pierre, HBC, shoots Columbia River rapids, 75
Lettuce, 38
Lewis, Morgan, 32
Lewis and Clark, plants collected, 25–26, 36, 41; travels, 40, 112, 155
Lewisia rediviva, 74
Lewiston, Idaho, 84–86
Licorice, wild, 63
Lily, pond, 24
glacier, 73
mariposa, 54, 184
Limnanthemum nymphoides, 22
Lindley, John, English botanist, 149, 157; criticized, 157–158; tests Douglas fir, 242; collection of Douglas's plants, 227n, 240n
Linnea borealis, 126
Linnean Society, London, admits Douglas, 147; Douglas's report to, 146–147, 177
Linton, George, HBC clerk, Fort George, Fraser River, 204
Little Wolf, Indian, 87
Liverpool, Earl of, 38
Liverpool Botanic Garden, 22
Liverpool Mercury, report of Douglas's death, 238
Lizards, at Fort Walla Walla, 168–169
Lobb, William, 245
Lockport, N.Y., 31
Lomi-lomi, aid in consuming poi, 218–219
London, Douglas's stay in, (1823), 19; (1827–29), 144–164
London *Morning Herald,* report of Douglas's death, 238
Long Expedition to Rocky Mountains (1819–20), 25
Lonicera hirsuta, 29
Lupinaster macrocephalus, 81
Lupine, 55, 185; golden, 81
Lupinus littoralis, 63
Sabinii, 81
Lycopodium, 24
Lyman, Mrs. D. B., quoted on Douglas's death, 235–237
Lyman, George D., 189

McDonald, Angus, describes Dalles des Morts, 110
McDonald, Archibald, HBC: accompanies Douglas to Snake River, 83, from Fort Nisqually to Fort Vancouver, 196; father of Ranald, 195; tribute by, 239
McDonald, Finan, HBC, 82–83, 120; mauled by buffalo, 122–123
MacDonald, Ranald, 195, 239

McDougall, George, HBC, makes ascent of Fraser River, 119
McGillivray, ——, Scone Palace gardens, 11–12
McGillivray's Rock, said to be Douglas's Mount Hooker, 138, 142
McKay, Jean Baptiste Desportes, HBC, on sugar pine search, 58, 71, 91–92, 97; cooks fine steaks, 94; a great hunter, 102
Mackenzie, Alexander, North West Company, 40
McKenzie, Alexander, HBC, 62
McKenzie, Donald, governor of Red River Colony, 125
McKinley, James (Santiago), California pioneer, 183
McLeod, Alexander Roderick, HBC, travels: Mackenzie River, 66; Umpqua River, 92–98, 102–103; HBC express, 107; Willamette and Santiam rivers, 169–170
McLeod, John, HBC, takes seeds across continent, 70, 72
McLoughlin, Dr. John, HBC superintendent, Columbia River region, 49; welcomes Douglas, 51; gives advice, 52, 92; kindness, 66, 69, 82, 102; awarded Horticultural Society's medal, 61; conducts matins, 66; his river gauge misunderstood by Indians, 54; express passenger, 107; on height of Athabaska Pass, 134–135; medical aides, 207–208
McMahon, Bernard, Philadelphia horticulturist, 26
McMillan, James, HBC, 105
McNeill, Captain William Henry, *Lama,* 193
Macrae, James, discoverer of silver sword plant, 216
McTavish, John George, HBC chief factor, York Factory, 127
Madeira, 43
Madrona, *see* Arbutus
Mansfield, Earl of, employs Douglas, 11; plants Douglas fir trees, 242
Manson, Donald, HBC, 93
Maple, broad-leaved, vine, 55–56; seeds, 90
Mariposa lily, *see* Lily
Markham, Violet R., quoted, 17, 103
Mauna Kea, 214–218, 226–234, 251–253
Mauna Loa, 222–231
Mauna Parrii, 166
Mentha borealis, 94
Mentzelia albicaulis, 55
  Lindleyii, 184–185
Menzies, Archibald, plants found: 40–42, 154; Douglas fir, 47, 154; gooseberry, 175; redwood tree, 177; seaweed, 196; climbs Mauna Loa, 226
Methow River, 73–75
Michaux, André, French botanist in Eastern America, 19–20, 39
Michigan, 30
Miles, ——, sawmill owner, Hawaii, 216
Mimulus Moschatus, 56–57
Mineral specimens, 108, 148; gold, 209–210; lead, etc., 210–211
Minerva, vessel, Honolulu to Hawaii, 232
Mint, Canada (plant), 94
Mission-bell, 71; bulbs, 108
Missionaries, Red River, 126; California, 187–188; Hawaii, 232–237
Monkey-flower, musk-scented, 56–57
Monterey, Calif., Douglas's arrival, 172; his residence, 175–184; his departure, 190
Monterey Company of Foreigners, 182–183
Moss, club, 24
Mount Brown, 130–143
Mount Diablo, California, 178
Mount Douglas, 143
Mount Hood, 51; height, 135; attempted ascents, 55, 208, 246
Mount Hooker, 130–143
Mount Jefferson, 51; height estimated, 169–170
Mount Rainier, Tolmie's partial ascent, 208
Mount St. Helens, 51, 196, 208
Mountains named after Douglas, 143
Multnomah, *see* Willamette
Munras, Estevan, California pioneer, 183

Munro, Donald, London gardener, spring and gooseberry named after, 84–85
Murray, John, London publisher, offers to publish Douglas's journal, 149
Murray, Stewart, superintendent, Glasgow Botanic Garden, 14, 18–19

Nechaco River, 201, 203, 204
Nelson, David, plant collector with Captain Cook, 40
Nemophila insignis, 176
Nesochen sandvicensis, 192
New Caledonia, 76; journey in, 197–206; route, 200–203; climate, 199; Douglas's prophecy for, 206
New Year's Day, dismal, 67
New York City, persons and places visited by Douglas, 23–27, 32–36
New York Botanical Garden, 24
New York Horticultural Society, Douglas attends meetings, 26, 32; Douglas becomes a member, 38
New York Statesman, 26
Niagara Falls, botanizing at, 31
Nicholas I, Emperor of Russia, member of Horticultural Society, 18, 148–149; sanctions Siberian journey, 198
Nicotiana, taken from Indian garden, 56
Nimrod, vessel, New York to London, 36
North Saskatchewan River, 122–123
North West Company, 40, 41, 128
Northwest America, *see* Douglas, David, Journeys
Northwest Passage, existence of, 66
Norway House, 123–124, 127
Nuphar, 24
Nuttall, Thomas, botanist, 20; ideals of, 33; criticized, 158; visit to Columbia River, 194–195, 245
Nymphaea, 24

Oaks, American, 39
Ocean-spray, 55
Oenothera, 55, 86, 184
Lindleyii, 157
speciosa, 38
triloba, 38
Ogden, Peter Skene, HBC, 89
Ohelo berry, offering to Pele, 220
Okanagan Lake, 201, 205, 210
Oliver, F. W., on the eminence of Robert Brown, 132
Olla-piska, Indian name for Douglas, 54–55
Orchideae, 44–45
Oregon boundary question, Douglas's opinion, 155–156
Oregon City, 58
Oregon grape, 55; tall, plant, 36; seeds, 90
Oregon sunshine, plant introduced, 244–245
Orontium, 24
Ortyx californica, 96
Otter, sea, 181
Ovis californianus, 148

Pacific Northwest, *see* Douglas, journeys
Pack rat, 78, 82
Paeonia Brownii, found, 81; lost, 121; named, 143; rediscovered, 168
Pambrun, Pierre, C., HBC, 76, 107
Pandanus Douglasii, use, 227
Parry, Dr. Charles Christopher, American botanist, rediscovery of *Abies amabilis*, 59–60
Pawquanaway, "Chiefess," 88
Paxton, Sir Joseph, 110n
Peale's Museum, Philadelphia, 26
Pedicularis canadensis, 38
Pele, goddess of volcanoes, 220–221
Pennsylvania, University of, Garden, 25, 33
Penstemon, 55, 75, 184–185
Perthshire Royal Horticultural Society, erects monument at Scone, 239–240
Phacelia, 184–185
Philadelphia, persons and places visited, 25–26, 33–34
Picea sitchensis, 60
Pickering, Dr. Charles, U.S. Exploring Expedition, investigates Douglas's death, 252

Pigeons, 31, 32; fight, 36; dissection, 38
Pine, big-cone or Coulter, 184
gray or Digger, 176–177
Monterey, 184
sugar, search for, 58, 71, 91–104; distance covered, 112; report on, 146–147; seeds, 145–146; as dessert, 92, 154
western white, 60
western yellow, 74, 191
Pine drops, 32, 58
Pinus amabilis, see *Abies*
Coulteri, 184
Douglasii, see Fir, Douglas
grandis, see *Abies*
insignis (*radiata*), 184
Lambertiana, see Pine, sugar
Menziesii, 60
monticola, 60
nobilis, see *Abies*
ponderosa, 74, 191
Sabiniana, 176–177
taxifolia, see Fir, Douglas
Pitcher plant, 24, 35
Plants, *see names and* APPENDIX B; Douglas's names retained, 56n; whereabouts of his herbaria, 227n, 240n
Platystemon californicus, 184–185
Plum, Washington, 24
Pogonia pendula, 38
Poi or poe, Hawaiian food, 218–219
Poppy, 185; California poppy, 154; bush poppy, 184–185
Portsmouth, England, 129
Preston, Sir Robert, Douglas employed by, 14
Priest's Rapids, 71
Primrose, evening, 38, 55, 86, 184; named after Lindley, 157
Prince, William and William Robert (son), nurserymen, Flushing, N.Y., differences with, 35–36, 39
Prince of Wales, vessel, Hudson Bay to England, 127, 129
Provencher, Rev. Joseph N., Red River Colony, 126
Pseudotsuga taxifolia, see Fir, Douglas
Pterospora andromedea, 32
Puget Sound, 196
Purisima Concepción Mission, La, 176n, 178
Pursh, Frederick, Austrian botanist, 20, 39
Pyrola, 37–38

Quail, 96
Quinine bush, 185

Rat, pack, 78, 82
Read, John, California pioneer, 213
Red Coat, Indian guide, 88
Red River Colony, persons and places visited, 124–126
Redwood tree, 177
Revelstoke, B.C., 109
Rhododendron, 58
Ribes irriguum, 85
sanguineum, 90, 146
speciosum, 175–176
Richardson, Dr. John, British naturalist, 38–39, 42; attends Finan McDonald, 123; his *Fauna Boreali-Americana* describes Douglas's zoological specimens, 153, 170; Richardson's plants described in Hooker's *Flora*, 132–133; on height of Rocky Mountains, 135, 141
Rickey's Rapids, 73, 75
Rio de Janeiro, 43–45
Robado, Spokane Indian guide, 87
Roberts, George B., HBC, 167; describes Douglas, 166
Robinson, Alfred, California pioneer, 189
Robinson Crusoe, 11, 45–46
Roche de Smet, height overestimated by Father de Smet, 136
Roche Miette, height overestimated by Aemilius Simpson, 139
Rochester, N.Y., 28, 31
Rocky Mountains, view of, 111; crossing of, 113; description of Mount Brown and Mount Hooker, 130–143, 246; plants of, 131, 141; height of, 134–136, 140
Rooke, Dr. T. C. R., postmortem on Douglas, 237

Roseburg, Ore., 102
Rowand, John, HBC chief factor, Fort Edmonton, 121–122, 127
Royal Geographic Society, on height of Rocky Mountains, 134–135
Royal Horticultural Society, places memorial tablet at Honolulu, 241. *See also* Horticultural Society of London
Royal Scottish Museum, Edinburgh, has Douglas's Franklin grouse, 116, 153–154
Royal Society, 161, 164
Russia, Douglas plants sent to, 186–187. *See also* Siberia

Sabbath observance, 12–13, 94, 126, 222–223
Sabine, Edward, secretary, Royal Society, 28; instructs Douglas, 161–165; reports on Douglas's work, 195–196, 240n; sorrows at his death, 238–239
Sabine, Joseph, secretary, Horticultural Society of London, 18, 23, 43, 147, 154; receives gifts from Clinton, 31; lupine named after Sabine, 81; pine, 176–177; satisfied with Douglas's work, 144–145; blamed by Douglas, 149–151; on Douglas's "fits," 162; resigns, 191–192
Sage, gray-ball, 185
Sagebrush, 71, 86
Sagittaria, 63
Salal, found by Menzies, 40; introduced by Douglas, 55, 154; berries stolen, 56; seeds obtained, 90; leaves used as a plate, 94; Douglas's favorite plant, 47, 127
Salmon, fishery, 57, 86, 87; price of, 57
Salvia carnosa, 185
San Antonio de Padua Mission, 176n, 178, 183
San Francisco, 176n, 178, 212–213
San Jose de Guadalupe Mission, 176n, 178
San Juan Bautista Mission, 176
San Luis Obispo Mission, 176n, 178
San Miguel Arcangel Mission, 176n, 178
San Rafael Mission, 176n, 178
Sandwich, Upper Canada, 30
Sandwich Islands, *see* Douglas, journeys
Santa Barbara, 44, 176n, 178
Santa Clara d'Assise Mission, 176
Santa Cruz Mission, 176, 177
Santa Lucia Mountains, 183
Santa Ynez Mission, 176n, 178
Santiam River, journeys on, (1826), 94, 104; (1830), 169, 170
Sarah and Elizabeth, vessel, Honolulu to England, 190
Sarcoramphos californica, see Condor
Sargent, Charles Sprague, American arboriculturist, on rediscovery of *Abies amabilis* and *Abies nobilis,* 59–60
Sarracenia purpurea, 24, 35
Sausalito, Calif., 213
Scarlet bugler, 184–185
Schachanaway, Chief, 65
Scone, Perthshire, 7–13; palace, 11–13; monument, 239–240
Scotland, Douglas's youth in, 7–16
Scott, ——, Douglas's friend, 12
Scouler, Dr. John, British naturalist, 42–48, 52, 154; penstemon named after him, 75; bones sent to him, 181
Screwpine, 227
Seaweed, 45, 196
Seneca Turnpike, 28
Sequoia sempervirens, 177
Sheep, mountain, 74, 94, 108, 118, 148
Shooting of birds on the wing, 53
Short, Rev. Patrick, California missionary, 189
Siberia, Douglas's proposed journey, 197–207
Side-saddle flower, 24, 35
Silver Mountain, 60
Silver sword plant, 215–216
Simpson, Lt. Aemilius, HBC, introduces apple trees, 105–106; on height of Rocky Mountains, 139–141
Simpson, George, HBC governor, names Committee's Punch Bowl,

116–117; gets clothes for Douglas, 124, 127; on height of Rocky Mountains, 140
Sitka, 204
Skeena River, 203
Smet, Rev. Pierre Jean de, hardships, 114n–115n; height of Rocky Mountains, 136
Snake River, 83–86; "Lewis and Clarke's River," 112
Soledad Mission, 176n, 178
Somerville, Rev. Thomas, quoted, 15
Spalding, Rev. H. H., quoted, 89
Spear, Nathan, California pioneer, 183
Spence, David, California pioneer, 189
Spiraea, 55
Spokane House, 73–74, 86, 112
Spokane River, 72–73
Spring beauty, bulbs, 108
Spruce, Sitka, 60
Stanley, Lord, president, Linnean Society, 147, 154
Steamboat Rapid, 109–110
Stewart, Alexander, Valleyfield, Fifeshire, 14
Stuart, John, HBC, 119–120
Stuart River, 201, 203, 204
Sturgeon, 52
Stuyvesant, Peter, governor of New Amsterdam, 20–21
Sudworth, George B., quoted, 96n
Sugar pine, *see* Pine
Sugar Pine Mountain, 102
Superior, Lake Erie steamship, 28
Symons, Lt. T. W., quoted, 77n

Tair, Captain, *Ann Maria*, 22
Taro, Hawaiian plant, 218–219
Tea praised, 53, 81, 103
Tetrao Franklinii, 116, 153–154
Tha-a-muxi, Chief, 62–65
Thames River, Upper Canada, 30
Theft by guide, 30
Thimbleberry, 74
Thomas, David, Aurora, N.Y., 31
Thompson, David, North West Company, 40; height of Rocky Mountains, 134–136, 140
Thomson, James, Elerslie, N.Y., 27, 32, 36
Thorburn, Grant, New York seed merchant, 32
Thorington, Dr. J. Monroe, on Mount Brown and Mount Hooker, 138–143
Thotmes III, King of Egypt, 17
Tobacco plant, 56; Tha-a-muxi's use, 63; use as currency, 70, 77, 88, 92, 101
Tollgate, Ore., 80
Tolmie, William Fraser, HBC, 135, 207–208
Torrey and Gray, *Flora of North America*, 186
Townsend, James K., American naturalist, 243–245
Tracey, ——, Albany, N.Y., 31
Trichostem lanceolatum, 94
Trifolium, 81, 86
Triphora pendula, 38

Umatilla River, 81
Umbellularia californica, 95–96
Umpqua River, 96–103, 112
United States, botany of, 19–21; Douglas's work in, 19–39
United States Exploring Expedition (1838–42), 167, 252

Vaccinium sandivicensis, 220
Valleyfield, Fifeshire, 14
Vancouver, Captain George, 40
Van Rensselaer, Stephen, Albany, N.Y., 31–32
Veitch, Henry, London nurseryman, 43
Victoria, Manuel, governor of California, 174
Vignes, Jean Louis, California pioneer, 183
Vigors, Nicholas, secretary, Zoological Society of London, 158–159
Violet, dogtooth, 73, 108
Volcanoes, Hawaii, 214–231

Walla Walla River, 79–82
Wallace, Robert, drowned in Columbia River, 110n, 128
Wallich, Nathaniel, Calcutta, 152
Wedgwood, John, 18
Wenatchee, Wash., 71

Wheeler, Arthur O., 142–143
Whirlpool River, 117
Whisky jack, bird, 117
Whitbey Harbour, *see* Gray's Harbor
Wilder, Gardner K., editor, *Hilo Tribune,* 253n
Wilkes, Lt. Charles, U.S. Exploring Expedition, 167, 252
Willamette River, journeys: (1825), 51, 58, 111; (1826), 93; (1830), 169–170
Willapa Bay, 63
William and Ann, voyage (1824–25), 3, 41–48, 52, 60, 62; wrecked, 47
Wilson, ——, schoolmaster, 9
Wilson, James, *Illustrations of Zoology,* 153, 246
Wilson, William, New York horticulturist, 32
Winnipeg, 125–126
Winnipeg River, 124
Wintergreen, 37
Wolf, deified, 108
Wood River, 113–115
Woodlands, Philadelphia, 26
Work, John, HBC, 64, 196; preserves birds, 76; on Snake River, 83–86; procures skins, 108
Wrangel, Baron Ferdinand Petrovich von, governor of Alaska, 197–199
Wyeth, Nathaniel J., 194–195

Xerophyllum asphodeloides, 58

York Factory, 127–129
Young Wasp, Indian guide, 79–82

Zoological Society of London, admits Douglas, 147–148, 158–159; assists Douglas, 163–164; receives Sandwich Island geese, 192; buys skins, 240n; receives birds, 246

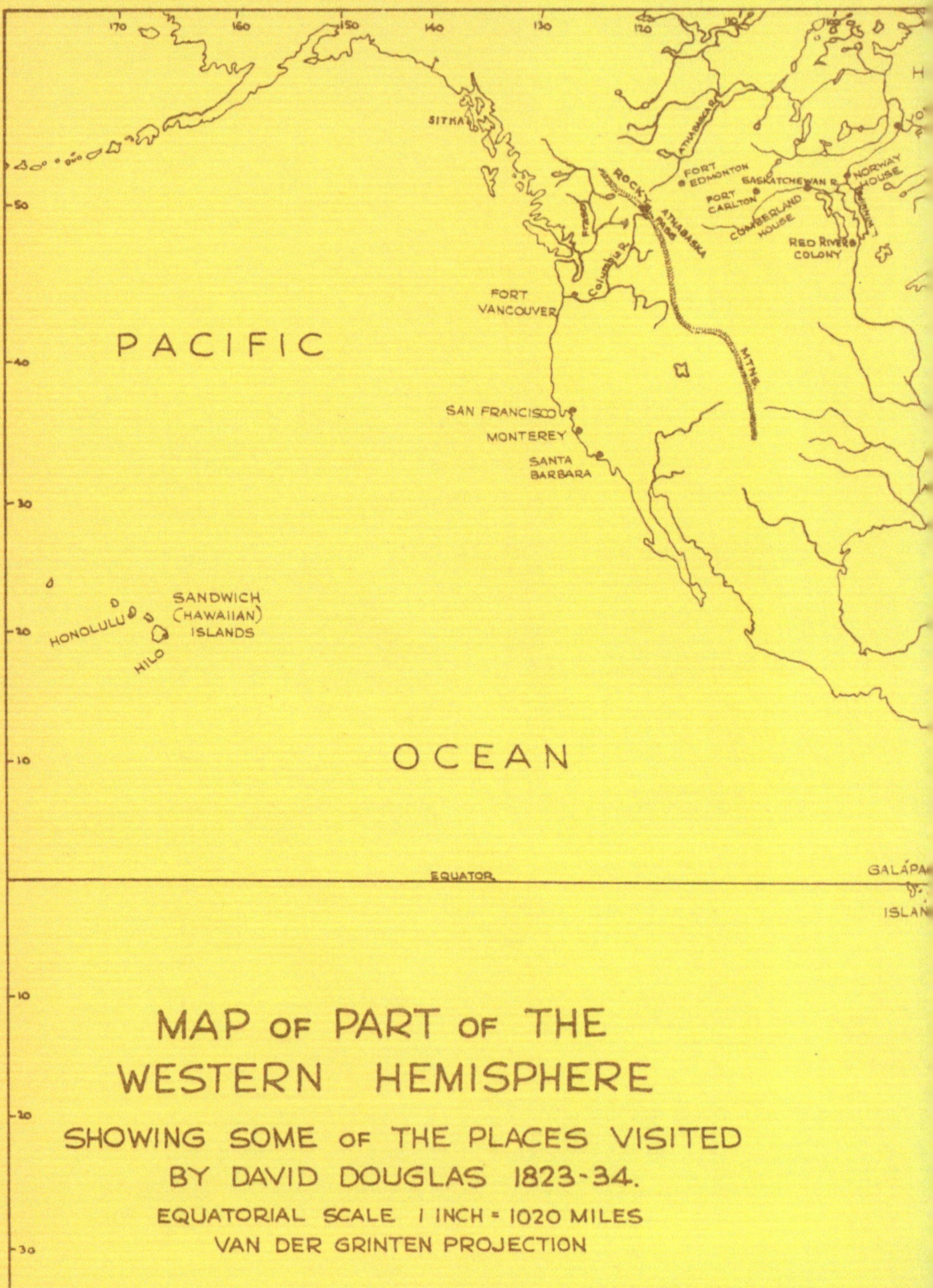

PACIFIC
OCEAN
SITKA
FORT EDMONTON
FORT CARLTON
SASKATCHEWAN R.
NORWAY HOUSE
CUMBERLAND HOUSE
RED RIVER COLONY
ROCKY
ATHABASKA PASS
ATHABASKA
ATHABASKA R.
Columbia R.
FORT VANCOUVER
MTNS.
SAN FRANCISCO
MONTEREY
SANTA BARBARA
SANDWICH (HAWAIIAN) ISLANDS
HONOLULU
HILO
EQUATOR
MAP OF PART OF THE WESTERN HEMISPHERE
SHOWING SOME OF THE PLACES VISITED BY DAVID DOUGLAS 1823-34.
EQUATORIAL SCALE 1 INCH = 1020 MILES
VAN DER GRINTEN PROJECTION

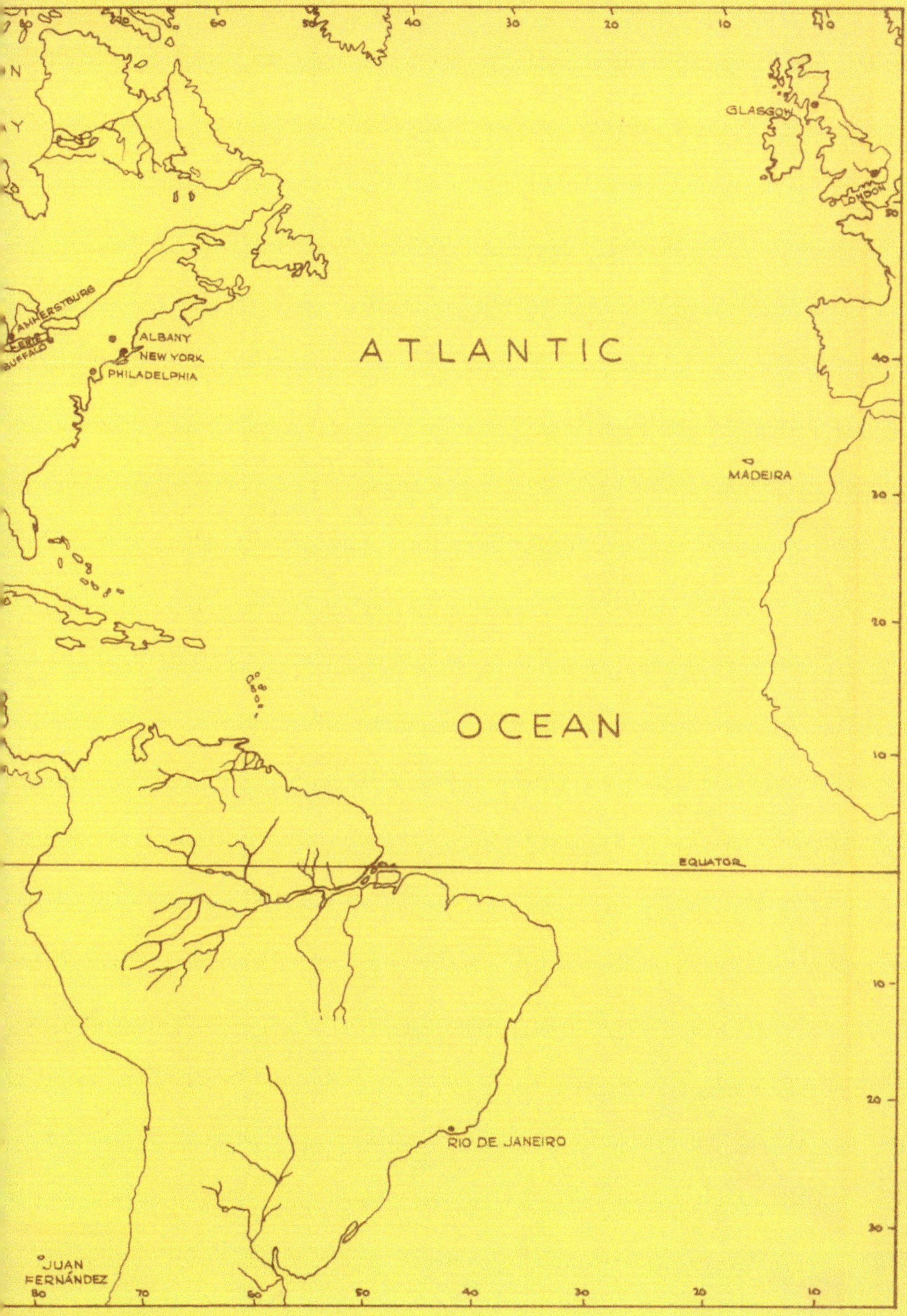

GLASGOW
LONDON
AMHERSTBURG
ERIE
BUFFALO
ALBANY
NEW YORK
PHILADELPHIA
ATLANTIC
OCEAN
MADEIRA
EQUATOR
RIO DE JANEIRO
JUAN FERNÁNDEZ

Bei Fragen zur Produktsicherheit wenden Sie sich bitte an:
If you have any questions regarding product safety,
please contact:

Walter de Gruyter GmbH
Genthiner Straße 13
10785 Berlin
productsafety@degruyterbrill.com